Aquifer Storage and Recovery in the Comprehensive Everglades Restoration Plan

A Critique of the Pilot Projects and Related Plans for ASR in the Lake Okeechobee and Western Hillsboro Areas

National Research Council
Division on Earth and Life Studies
Water Science and Technology Board
Board on Environmental Studies and Toxicology
Committee on Restoration of the Greater Everglades Ecosystem

NATIONAL ACADEMY PRESS
Washington, D.C.

NATIONAL ACADEMY PRESS • 2101 Constitution Avenue, N.W. • Washington, D.C. 20418

NOTICE: The project that is the subject of this report was approved by the Governing Board of the National Research Council, whose members are drawn from the councils of the National Academy of Sciences, the National Academy of Engineering, and the Institute of Medicine. The members of the committee responsible for the report were chosen for their special competencies and with regard for appropriate balance.

Supported by the South Florida Ecosystem Restoration Task Force, Department of Interior, under assistance of Cooperative Agreement No. 5280-9-9029. The views and conclusions contained in this document are those of the authors and should not be interpreted as necessarily representing the official policies, either expressed or implied, of the Government.

This report is available from the National Academy Press, 2101 Constitution Avenue, N.W., Washington, D.C. 20418, (800) 624-6242 or (202) 334-3313 (in the Washington metropolitan area); internet <http://www.nap.edu>.

International Standard Book Number 0-309-07347-2

Copyright 2001 by the National Academy of Sciences. All rights reserved.

THE NATIONAL ACADEMIES
Advisers to the Nation on Science, Engineering, and Medicine

National Academy of Sciences
National Academy of Engineering
Institute of Medicine
National Research Council

Printed in the United States of America

The **National Academy of Sciences** is a private, nonprofit, self-perpetuating society of distinguished scholars engaged in scientific and engineering research, dedicated to the furtherance of science and technology and to their use for the general welfare. Upon the authority of the charter granted to it by the Congress in 1863, the Academy has a mandate that requires it to advise the federal government on scientific and technical matters. Dr. Bruce M. Alberts is president of the National Academy of Sciences.

The **National Academy of Engineering** was established in 1964, under the charter of the National Academy of Sciences, as a parallel organization of outstanding engineers. It is autonomous in its administration and in the selection of its members, sharing with the National Academy of Sciences the responsibility for advising the federal government. The National Academy of Engineering also sponsors engineering programs aimed at meeting national needs, encourages education and research, and recognizes the superior achievement of engineers. Dr. William A. Wulf is president of the National Academy of Engineering.

The **Institute of Medicine** was established in 1970 by the National Academy of Sciences to secure the services of eminent members of appropriate professions in the examination of policy matters pertaining to the health of the public. The Institute acts under the responsibility given to the National Academy of Sciences by its congressional charter to be an adviser to the federal government and, upon its own initiative, to identify issues of medical care, research, and education. Dr. Kenneth I. Shine is president of the Institute of Medicine.

The **National Research Council** was organized by the National Academy of Sciences in 1916 to associate the broad community of science and technology with the Academy's purposes of 0furthering knowledge and advising the federal government. Functioning in accordance with general policies determined by the Academy, the Council has become the principal operating agency of both the National Academy of Sciences and the National Academy of Engineering in providing services to the government, the public, and the scientific and engineering communities. The Council is administered jointly by both Academies and the Institute of Medicine. Dr. Bruce M. Alberts and Dr. William A. Wulf are chairman and vice chairman, respectively, of the National Research Council.

COMMITTEE ON RESTORATION OF THE GREATER EVERGLADES ECOSYSTEM

JAMES M. DAVIDSON, *Chair*, University of Florida (ret.), Gainesville
SCOTT W. NIXON, University of Rhode Island, Narragansett
JOHN S. ADAMS, University of Minnesota, Minneapolis
JEAN M. BAHR, University of Wisconsin, Madison
LINDA K. BLUM, University of Virginia, Charlottesville
PATRICK L. BREZONIK, University of Minnesota, St. Paul
FRANK W. DAVIS, University of California, Santa Barbara
WAYNE C. HUBER, Oregon State University, Corvallis
STEPHEN R. HUMPHREY, University of Florida, Gainesville
DANIEL P. LOUCKS, Cornell University, Ithaca, New York
GORDON H. ORIANS, University of Washington, Seattle (resigned December 2000)
KENNETH W. POTTER, University of Wisconsin, Madison
LARRY ROBINSON, Florida Agricultural and Mechanical University, Tallahassee
STEVEN E. SANDERSON, Emory University, Atlanta, Georgia
REBECCA R. SHARITZ, Savannah River Ecology Laboratory, Aiken, South Carolina, and University of Georgia, Atlanta
JOHN VECCHIOLI, U.S. Geological Survey (ret.), Tallahassee, Florida

NRC Staff

STEPHEN D. PARKER, Director, Water Science and Technology Board
DAVID J. POLICANSKY, Associate Director, Board on Environmental Studies and Toxicology
WILLIAM S. LOGAN, Staff Officer, Water Science and Technology Board
PATRICIA JONES KERSHAW, Staff Associate, Water Science and Technology Board

WATER SCIENCE AND TECHNOLOGY BOARD

HENRY J. VAUX, Jr., *Chair*, Division of Agriculture and Natural Resources, University of California, Oakland
RICHARD G. LUTHY, *Vice Chair*, Carnegie Mellon University, Pittsburgh, Pennsylvania
RICHELLE M. ALLEN-KING, Washington State University, Pullman
GREGORY B. BAECHER, University of Maryland, College Park
JOHN BRISCOE, The World Bank, Washington, D.C.
EFI FOUFOULA-GEORGIOU, University of Minnesota, Minneapolis
STEVEN P. GLOSS, University of Wyoming, Laramie
WILLIAM A. JURY, University of California, Riverside
GARY S. LOGSDON, Black & Veatch, Cincinnati, Ohio
DIANE M. MCKNIGHT, University of Colorado, Boulder
JOHN W. MORRIS, J.W. Morris Ltd., Arlington, Virginia
PHILIP A. PALMER (Retired), E.I. du Pont de Nemours & Co., Wilmington, Delaware
REBECCA T. PARKIN, The George Washington University, Washington, D.C.
RUTHERFORD H. PLATT, University of Massachusetts, Amherst
JOAN B. ROSE, University of South Florida, St. Petersburg
JERALD L. SCHNOOR, University of Iowa, Iowa City
R. RHODES TRUSSELL, Montgomery Watson, Pasadena, California

Staff

STEPHEN D. PARKER, Director
LAURA J. EHLERS, Senior Staff Officer
CHRIS ELFRING, Senior Staff Officer
JEFFREY W. JACOBS, Senior Staff Officer
MARK C. GIBSON, Staff Officer
WILLIAM S. LOGAN, Staff Officer
M. JEANNE AQUILINO, Administrative Associate
PATRICIA JONES KERSHAW, Study/Research Associate
ANITA A. HALL, Administrative Assistant
ELLEN DE GUZMAN, Senior Project Assistant
ANIKE L. JOHNSON, Project Assistant

BOARD ON ENVIRONMENTAL STUDIES AND TOXICOLOGY

GORDON ORIANS (*Chair*), University of Washington, Seattle, Washington
JOHN DOULL, University of Kansas Medical Center, Kansas City, Kansas
DAVID ALLEN, University of Texas, Austin, Texas
INGRID C. BURKE, Colorado State University, Fort Collins, Colorado
THOMAS BURKE, Johns Hopkins University, Baltimore, Maryland
GLEN R. CASS, Georgia Institute of Technology, Atlanta, Georgia
WILLIAM L. CHAMEIDES, Georgia Institute of Technology, Atlanta, Georgia
CHRISTOPHER B. FIELD, Carnegie Institute of Washington, Stanford, California
JOHN GERHART, University of California, Berkeley, California
J. PAUL GILMAN, Celera Genomics, Rockville, Maryland
DANIEL S. GREENBAUM, Health Effects Institute, Cambridge, Massachusetts
BRUCE D. HAMMOCK, University of California, Davis, California
ROGENE HENDERSON, Lovelace Respiratory Research Institute, Albuquerque, New Mexico
CAROL HENRY, American Chemistry Council, Arlington, Virginia
ROBERT HUGGETT, Michigan State University, East Lansing, Michigan
JAMES F. KITCHELL, University of Wisconsin, Madison, Wisconsin
DANIEL KREWSKI, University of Ottawa, Ottawa, Ontario
JAMES A. MACMAHON, Utah State University, Logan, Utah
CHARLES O'MELIA, Johns Hopkins University, Baltimore, Maryland
WILLEM F. PASSCHIER, Health Council of the Netherlands, The Hague
ANN POWERS, Pace University School of Law, White Plains, New York
KIRK SMITH, University of California, Berkeley, California
TERRY F. YOSIE, American Chemistry Council, Arlington, Virginia

Senior Staff

JAMES J. REISA, Director
DAVID J. POLICANSKY, Associate Director and Senior Program Director for Applied Ecology
RAYMOND A. WASSEL, Senior Program Director for Environmental Sciences and Engineering
KULBIR BAKSHI, Program Director for the Committee on Toxicology
ROBERTA M. WEDGE, Program Director for Risk Analysis
K. JOHN HOLMES, Senior Staff Officer

Preface

This report is a product of the Committee on Restoration of the Greater Everglades Ecosystem (CROGEE), which provides consensus advice to the South Florida Ecosystem Restoration Task Force ("Task Force"). The Task Force was first established in 1993 and was codified in the 1996 Water Resources Development Act (WRDA); its responsibilities include the development of a comprehensive plan for restoring, preserving and protecting the South Florida ecosystem, and the coordination of related research. The CROGEE works under the auspices of the Water Science and Technology Board and the Board on Environmental Studies and Toxicology of the National Research Council.

Much, but not all, of the material used for this report came directly or indirectly from the Aquifer Storage and Recovery (ASR) workshop held by the CROGEE in Miami, Florida, on October 19, 2000. The workshop was open to the public and was attended by about 60 people including personnel from the South Florida Water Management District (SFWMD) and U.S. Army Corps of Engineers (USACE), federal, state and local agencies, universities, consulting firms, and environmental organizations. There were 10 invited experts from government, academia and the private sector, and eight members of the CROGEE present. The workshop agenda and list of participants are shown in Appendices A and B, respectively.

ASR was chosen as the workshop topic for several important reasons. First, ASR is a critical element of the restoration effort, accounting for a significant amount of the storage of water presently being lost to tide through the St. Lucie and Caloosahatchee basins. The restoration plan anticipates capturing this water during wet intervals and releasing it for ecological use and for water supply when needed.

Second, due to the technical and regulatory uncertainties of implementing ASR at the proposed scale, the 1999 WRDA contained authorization for the execution of two ASR pilot projects – one for the area around Lake Okeechobee, and another for the western Hillsboro site in Palm Beach County. These two pilot projects were the first of the 31 projects and six pilot projects to be implemented under the Master Project Management Plan to have individual Project Management Plans prepared for them. Thus, there were written materials available for the committee to evaluate.

Finally, these documents were still in a draft form, and their authors have been actively soliciting input from federal, state, and local agencies, academia, and the private sector. The designers of the Comprehensive Everglades Restoration Plan (CERP) have incorporated into the plan the concept of *adaptive assessment*, i.e., the development of a protocol for collecting and interpreting new information for the purpose of improving the design of the restoration plan. The two ASR pilot projects seemed to be highly appropriate targets for the application of this principle; that is, the time appeared right to help the USACE and SFWMD to maximize the opportunities for learning from the ASR pilot projects.

Thus, the ASR pilot projects were chosen for attention by the CROGEE not because they were perceived to be any more or less well designed than other aspects of the Comprehensive Plan, but because of their importance to the overall restoration effort, and because they were at an optimal planning stage for constructive feedback.

The primary documents submitted to the committee for their evaluation were the "second drafts" of the Project Management Plans for the Lake Okeechobee and Western Hillsboro Aquifer

Storage and Recovery Pilot Projects, dated September 2000. (Later drafts of these plans now are available but were not evaluated by the committee for this report.) Following receipt of these documents, but prior to the workshop, the committee submitted a list of questions about the plans, organized by theme, to the designers of the pilot projects. The answers to these questions were supplied to the CROGEE several days before the workshop (Appendix C). This procedure saved valuable time during the workshop, and enabled the committee to focus on specific issues of interest.

The workshop was planned with the cooperation and assistance of Peter Kwiatkowski, Project Manager (SFWMD) for the Lake Okeechobee ASR Pilot Project; Rick Nevulis, Project Manager (SFWMD) for the Western Hillsboro ASR Pilot Project; Glenn Landers, Project Manager (USACE) for both Pilot Projects; Terrence "Rock" Salt, Executive Director of the Task Force; Peter Ortner, liaison to the South Florida Ecosystem Restoration Working Group; Paul Dresler, liaison to the Department of the Interior; and many others. Peter Kwiatkowski also took the lead for the Task Force agencies to present the ASR plans at the workshop.

The Committee is grateful for the participation in the workshop of the following invited experts: Walt Schmidt, Florida Geological Survey; Robert Renken, U.S. Geological Survey; Donald McNeill, University of Miami; James Cowart, Florida State University; Joan Rose, University of South Florida; Richard Harvey, U.S. Environmental Protection Agency; Joan Browder, Southeast Fisheries Science Center; Mark Pearce, Water Resource Solutions Inc.; Rich Deuerling, Florida Department of Environmental Protection; and Tom Missimer, CDM-Missimer International. We are also grateful for the oral and written comments of other participants in the workshop. The Committee also acknowledges the previous efforts of the ASR Issue Team members, many of whom attended this workshop, who identified many of the crucial gaps in knowledge in their 1999 report to the Task Force.

Within the CROGEE, the Workshop was planned by a subgroup consisting of CROGEE members Jean Bahr, Patrick Brezonik, and John Vecchioli and NRC staff officer Will Logan. Following the workshop, the CROGEE members present met to outline their major conclusions. These were drafted, presented to the full CROGEE, and revised during a closed meeting the following day. Subsequent to the workshop, participants submitted additional materials to the committee (Appendix D), and these were also considered in drafting the report. All members have had a subsequent opportunity to review the full text of the report, which should be considered as a consensus report of the full CROGEE. It should also be emphasized that these conclusions do not necessarily reflect the opinions of the other participants in the workshop.

This report has been reviewed in draft form by individuals chosen for their diverse perspectives and technical expertise, in accordance with procedures approved by the NRC's Report Review Committee. The purpose of this independent review is to provide candid and critical comments that will assist the institution in making its published report as sound as possible and to ensure that the report meets institutional standards for objectivity, evidence, and responsiveness to the study charge. The review comments and draft manuscript remain confidential to protect the integrity of the deliberative process. We wish to thank the following individuals for their review of this report:

   Wilfried H. Brutsaert, Cornell University, Ithaca, New York
   James Crook, Black & Veatch, Boston, Massachusetts
   James T. Morris, University of South Carolina, Columbia
   Zhuping Sheng, El Paso Water Utilities, Texas
   Carol Wicks, University of Missouri, Columbia
   William W. Woessner, University of Montana, Missoula

Although the reviewers listed above have provided many constructive comments and suggestions, they were not asked to endorse the conclusions or recommendations nor did they see the final draft of the report before its release. The review of this report was overseen by George M. Hornberger, University of Virginia, appointed by the Division on Earth and Life Studies. Dr. Hornberger was responsible for making certain that an independent examination of this report was carried out in accordance with institutional procedures and that all review comments were carefully considered. Responsibility for the final content of this report rests entirely with the authoring committee and the institution.

James M. Davidson, Chair
Committee on Restoration of the Greater Everglades Ecosystem

Jean M. Bahr, Chair
CROGEE Subcommittee on Aquifer Storage and Recovery

Contents

EXECUTIVE SUMMARY 1
    Concerns about the role of ASR in the CERP _____ 1
    The pilot projects _____ 1
    The Committee's Charge _____ 2
    Regional Science Issues _____ 3
    Water Quality Issues _____ 3
    Local Performance/Feasibility Issues _____ 4
    General Conclusions _____ 4

INTRODUCTION 6
    Importance of ASR to Comprehensive Everglades Restoration Plan (CERP) _____ 6
    Concerns expressed about large-scale ASR in South Florida _____ 7
    The Lake Okeechobee and Western Hillsboro ASR Projects _____ 8
    The Committee's Charge _____ 10

REGIONAL SCIENCE ISSUES 12
    Motivation _____ 12
    Issues Discussed _____ 12
    Conclusions and Recommendations _____ 13

WATER QUALITY ISSUES 15
    Motivation _____ 15
    Issues Discussed _____ 15
    Conclusions and Recommendations _____ 16

LOCAL PERFORMANCE/FEASIBILITY ISSUES 18
    Motivation _____ 18
    Issues Discussed _____ 18
    Conclusions and Recommendations _____ 19

SUMMARY AND CONCLUSIONS 21

REFERENCES 23

APPENDIX A WORKSHOP AGENDA 26

APPENDIX B WORKSHOP PARTICIPANTS 28

APPENDIX C QUESTIONS SENT TO THE SFWMD PRIOR TO THE WORKSHOP, AND ITS RESPONSES _____ 30

APPENDIX D WORKSHOP-RELATED MATERIALS RECEIVED BY THE COMMITTEE AFTER THE WORKSHOP AND PRIOR TO FINALIZATION OF THE REPORT _____ 40

APPENDIX E EXCERPTS FROM DRAFT PROJECT MANAGEMENT PLAN – LAKE OKEECHOBEE _____ 41

APPENDIX F EXCERPTS FROM DRAFT PROJECT MANAGEMENT PLAN – WESTERN HILLSBORO _____ 46

APPENDIX G BIOGRAPHICAL SKETCHES OF COMMITTEE MEMBERS _____ 55

Aquifer Storage and Recovery in the Comprehensive Everglades Restoration Plan

A Critique of the Pilot Projects and Related Plans for ASR in the Lake Okeechobee and Western Hillsboro Areas

Executive Summary

Aquifer storage and recovery (ASR) is a process by which water is recharged through wells to an aquifer and extracted for beneficial use at some later time from the same wells. ASR is proposed as a major water storage component in the Comprehensive Everglades Restoration Plan (CERP), developed jointly by the U.S. Army Corps of Engineers (USACE) and the South Florida Water Management District (SFWMD). The plan would use the Upper Floridan aquifer (UFA) to store as much as 1.7 billion gallons per day (gpd) (6.3 million m^3/day) of excess surface water and shallow groundwater during wet periods for recovery during seasonal or longer-term dry periods, using about 333 wells. ASR represents about one-fifth of the total estimated cost of the CERP.

ASR may have advantages over surface storage in south Florida in that it may limit losses due to evaporation when compared with surface storage and limit the acreage of land removed from other productive uses. It also may permit the recovery of large volumes of water during severe, multi-year droughts to augment deficient surface water supplies.

Concerns about the role of ASR in the CERP

Concerns have been voiced about the use of ASR. Some of these are related to its proposed scale, which is much larger than existing ASR projects in Florida. Regional issues include the relative scarcity of subsurface information in areas where ASR wells will be located, and impacts of the combined hydraulic head increases from the regional scale ASR on existing ASR wells, supply wells, and underground injection control (UIC) monitoring wells. Water quality issues include the suitability of the source waters for recharge without extensive pretreatment, likely water quality changes during storage in the aquifer, and whether the quality of the recovered water will pose environmental or health concerns. Local performance issues include whether the proposed ASR injection volumes will result in pressures sufficient to cause rock fracturing, and lack of information concerning the relationships among ASR storage zone properties, recovery rates, and recharge volumes.

The pilot projects

To address some of these issues, the CERP proposed several ASR pilot projects, two of which were approved in the 1999 Water Resources Development Act (WRDA) and are the subject of this report.

Lake Okeechobee. According to the draft project management plan (PMP), this pilot project is designed to test the feasibility of placing about 200 ASR wells with an estimated capacity of one billion gpd (3.8 million m^3/day) near Lake Okeechobee. The full-scale CERP feature is designed to (1) provide regional storage, (2) increase the lake's water storage capability, (3) better manage regulatory releases from the lake; (4) reduce harmful discharges to the St. Lucie and Caloosahatchee estuaries; and (5) enhance flood protection.

The stated purposes of the pilot project are to (1) install several exploratory/test ASR systems in geographically dispersed areas around the lake, (2) determine the water quality characteristics of waters to be recharged, water recovered from the aquifer, and the water in the receiving aquifer, (3) provide an estimate of the amount of water that can be recharged, and (4) provide hydrogeological and geotechnical information on the UFA within the region, and the ability of the UFA to maintain injected water for future recovery.

Western Hillsboro. This CERP feature involves a series of ASR wells that would be located next to an above-ground reservoir in Palm Beach County, or along the Hillsboro Canal. The ASR capacity would be about 150 million gpd (570,000 m^3/day), from 30 wells pumping at 5 million gpd (19,000 m^3/day) per well. Surface water and/or groundwater from the surficial aquifer adjacent to the reservoir would be the source of recharge water. This feature is designed to supplement water deliveries to the Hillsboro Canal during dry periods, thereby reducing demands on Lake Okeechobee and the Loxahatchee National Wildlife Refuge. Water would be pumped into the UFA during times of excess, and returned to the canal to help maintain canal stages during the times of deficit.

The stated purposes of the pilot project are (1) to determine the most suitable sites and the optimum configuration for the ASR wells in the vicinity of the reservoir, (2) to evaluate many of the hydrogeological and geotechnical characteristics of the soils and aquifer in the area, and (3) to determine the specific water quality characteristics of water within the aquifer, water proposed for recharge, and water recovered from the aquifer.

The Committee's Charge

The National Research Council's Committee on Restoration of the Greater Everglades Ecosystem (CROGEE) examined the second draft of the pilot project PMPs from the perspective of adaptive assessment, i.e., the extent to which the pilot projects will contribute to process understanding that can improve design and implementation of restoration project components. It organized a workshop in Miami, Florida on October 19, 2000 to discuss these plans with SFWMD and USACE personnel and other interested parties. Immediately prior to the workshop, much of the proposed work on regional analysis of the subsurface was extracted from the pilot projects and reorganized into a planned, but as yet unfunded, regional ASR study. Because the CROGEE concludes that this regional work is crucial to evaluating the potential for success of the ASR elements of the CERP, this report is a critique of the pilot projects *(sensu stricto) and related studies*.

This report does not make judgements regarding the overall desirability of ASR as a major component of the CERP, either in absolute terms or in comparison with other storage options such as surface reservoir storage.

The committee's recommendations are organized into three categories—regional science issues, water quality issues, and local performance/feasibility issues—and are as follows:

Regional Science Issues

The scale of the proposed ASR projects is unprecedented, with approximately 1.7 billion gpd (6.3 million m^3/day) concentrated in a relatively small area compared with the current total withdrawal of about 3 billion gpd (11 million m^3/day) that is pumped regionally from the UFA. It is critical to assess the aggregate hydraulic effects of ASR wells on the existing UFA flow system. This will require a regional hydrogeologic assessment and interpretation of the ASR pilot projects within a regional context. Recently developed plans to pursue funding for a regional study beyond the constraints of the pilot projects are commendable. Essential elements of the regional study should include the following.

- Development of a preliminary list of data needs and compilation of available data for a regional assessment. This task should be undertaken as soon as possible.
- Development of a regional-scale groundwater flow model. Model development should proceed in parallel with initial data compilation, and can be used to identify data gaps.
- Drilling of exploratory wells in key areas, including core sampling, downhole geophysical logging, hydraulic testing and water quality sampling of these wells.
- Seismic reflection surveys, used in conjunction with results from exploratory wells, to constrain the three-dimensional geometry and continuity of hydrostratigraphic units.
- Use of the regional model in conjunction with other regional data sets to develop a rational, multi-objective approach to ASR facility siting during final design of the regional ASR systems.

Water Quality Issues

ASR water may be used for agriculture, for augmenting water inflows to natural ecosystems in the Everglades, and, indirectly, to supplement municipal (drinking water) supplies. Different regulations and different concerns about water quality arise in connection with these different intended uses. Thus, considerations must be broader than simply meeting existing water quality criteria. The chemical analyses planned as part of the pilot projects are insufficient to answer many questions about potential biological impacts of ASR water on the Everglades ecosystem. Such analyses, by themselves, also cannot provide the mechanistic information needed to develop geochemical models to predict how water chemistry/water quality will change in a full-scale ASR program. Water quality studies related to ASR should be expanded to include the following.

- Scientific studies, including laboratory and field bioassays and ecotoxicological studies, to help determine appropriate standards that consider not only the initial receptors of the recovered water, but also downstream receptors.
- Characterization of organic carbon of the source water and studies designed to anticipate the effects of this material on biogeochemical processes in the subsurface. These are a high priority because recharge water is likely to have high concentrations of dissolved organic carbon compared with ambient water in the aquifer.
- Laboratory studies to evaluate dissolution kinetics and redox processes that could release major ions, heavy metals, arsenic, radionuclides, and other constituents from the aquifer matrix. These should be undertaken before (and/or in parallel with) water quality monitoring during cycle testing.

- Studies designed to enhance understanding of mechanisms responsible for mixing of dilute recharge water with brackish to saline groundwater. These are necessary to predict changes in dissolved solids due to mixing.

Local Performance/Feasibility Issues

In keeping with the principle of adaptive assessment, the CROGEE recommends that the pilot projects not be considered simply as demonstrations at particular sites. Instead, they should be viewed as an opportunity to develop a better understanding of the hydrogeologic and well construction characteristics that control the relationships between storage intervals, recharge volumes, and recoverability. Therefore, the pilot projects should be designed to maximize the value of the data obtained for improving understanding of ASR performance at both the specific sites tested and more generally in the UFA. In case of budget constraints, it would be preferable to do more detailed studies with enhanced monitoring at a reduced number of sites or with a reduced number of ASR wells, rather than more limited studies using the currently planned number of exploratory wells. Important elements that should be considered in design of the pilot tests are listed below.

- Monitoring wells that allow sampling at discrete depths and in a variety of directions from the recharge well. These should be a high priority at each pilot ASR site for purposes of delineating bubble geometry, potential preferential flow paths, and the extent of mixing between recharged and ambient water.
- Tests designed to compare effects of short and long sections of open borehole on well performance. These would be particularly useful for optimizing design of ASR wells.
- Continued recharge of the Lake Okeechobee pilot wells for periods on the order of a year or more during cycle testing. Hydrologic models indicate that continuous recharge is likely to be required for multi-year periods. Effects of mixing and water-rock interactions over these time periods may not be readily predicted on the basis of short-term cycle results.
- Use of observations obtained from a suitable network of monitoring wells, and over suitable time periods, to develop new or improved conceptual and numerical models of "bubble" migration. Such models may need to account for three-dimensional, density dependent flow and solute transport in multi-permeability media containing fractures and solution conduits.

General Conclusions

An improved understanding in these three areas of uncertainty – regional science, water quality, and local feasibility – will require studies that go beyond the scope of the proposed ASR pilot projects, which are almost exclusively devoted to local feasibility issues. A regional study and modeling effort recently was proposed as a separate project, but has not been authorized or funded. Ideally, this would *precede* any local-scale feasibility studies to allow these studies to make use of regional information. The pilot projects are also lacking studies to help assess the appropriate water quality standards for discharge of recovered water to ecosystems.

Information from the regional study, water quality studies, and local scale pilot studies ultimately must be synthesized and used together in the overall assessment of ASR as a component of the CERP. With an improved understanding of ASR systems that will result from these studies, it should be possible to develop ASR system designs that then can be compared to other storage options in terms of overall performance, including storage capacity and cost effectiveness. This will require further examination of related issues that were not the focus of the workshop, but are important to the overall assessment of ASR feasibility and effectiveness. These include performance of surface and subsurface storage, a comparison of losses from surface evaporation and subsurface storage, the number and distribution of wells required for a regional ASR system, and estimates of energy costs for operation over the anticipated project life.

As a final comment, the CROGEE notes that the CERP calls for ASR to be implemented in phases. The Committee agrees that phased implementation is an appropriate strategy and strongly recommends a) thorough evaluation of the environmental effects of each incremental increase in scale of ASR, and b) ongoing adaptive assessment of the program.

1

Introduction

Artificial recharge, i.e., the process by which water is directed into the ground to replenish an aquifer (NRC, 1994), has taken on increasing importance in recent decades. Aquifer storage and recovery (ASR), which is the major theme of this report, may be viewed as a sub-field within this more general area. In this report, ASR refers to, "the storage of water in a suitable aquifer through a well during times when water is available, and recovery of the water from the same well during times when it is needed" (Pyne, 1995). This is shown conceptually in Figure 1. The advantages of this approach over other methods of artificial recharge may include reduced problems with plugging, economic benefits from the efficient use of the wells, and the potential to use aquifers with less than optimal water-quality characteristics (Pyne, 1995).

Importance of ASR to Comprehensive Everglades Restoration Plan (CERP)

ASR is proposed as a major water storage component in the Comprehensive Everglades Restoration Plan (CERP), developed jointly by the U.S. Army Corps of Engineers (USACE) and the South Florida Water Management District (SFWMD). The ASR proposal would use porous and permeable units in the Upper Floridan aquifer (UFA) to store excess surface water and shallow groundwater at rates of up to 1.7 billion gallons per day (gpd) (6.3 million m^3/day) during wet periods for recovery during seasonal or longer-term dry periods (USACE, 1999; SFWMD, 2000). Based on the 1965-1995 record, average and maximum yearly recharge volumes in the Lake Okeechobee area are estimated to be about 264,000 acre-ft (86 billion gallons; 326 million m^3) and 1,100,000 acre-ft (358 billion gallons; 1.36 billion m^3) per year, respectively (USACE, 1999). The latter would require pumping all 200 proposed ASR wells in that region at capacity year round. Average yearly recovery volumes are estimated to be somewhat lower - about 136,000 acre-ft (44 billion gallons; 168 million m^3) per year (USACE, 1999). ASR represents about one-fifth of the total cost of the CERP (Kwiatkowski, 2000a).

The CERP suggests that ASR may provide greater storage efficiency when compared to land requirements and high seepage and evapotranspiration rates associated with above ground reservoir storage (USACE, 1999). ASR may offer particular advantages over surface storage in South Florida where land acquisition costs are high and flat topography coupled with a shallow water table place constraints on surface reservoir construction. Additional advantages cited for this strategy are that ASR wells can be located in areas of greatest need, thus reducing water distribution costs, and that ASR permits recovery of large volumes of water during severe, multi-year droughts to augment deficient surface water supplies. ASR may also limit certain kinds of degradation (e.g., algal blooms, nutrient and pathogen inputs from birds, etc.) that may occur with surface storage. Potential disadvantages include losses to the aquifer due to mixing within saline aquifers, and low recharge and recovery rates relative to surface storage (USACE, 1999, Appendix B).

The ASR concept in South Florida involves the recharge of excess fresh surface and shallow groundwater during wet periods into the UFA through approximately 333 wells. This assumes recharge rates of about five million gpd (19,000 m^3/day) per well; the exact number of wells would be a function of their long-term capacity. Ambient groundwater in the UFA is brackish to saline

(CH2M Hill, 1989). Because of this high salinity, the UFA is currently used little in these areas for water supply. During the recharge phase of ASR system operation, the ambient groundwater would be displaced by the injected fresh water such that a zone, or "bubble", of fresh water would be created and stored around each ASR well. This bubble of fresh water could be drawn upon later by the same ASR wells and the recovered water used to augment deficient surface water supplies during dry seasons or longer-term drought periods. In essence, ASR would use subsurface space in the UFA as the reservoir for storing water.

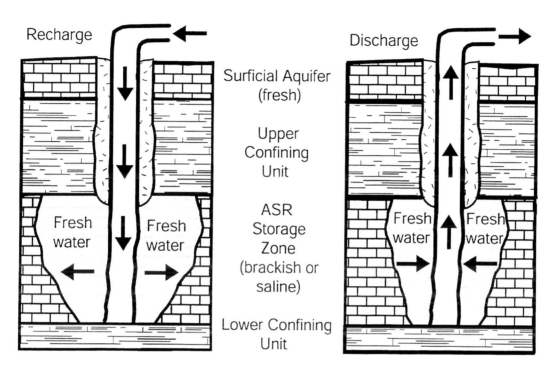

FIGURE 1. Schematic diagram of the recharge and recovery phases of ASR for a typical south Florida system. The relatively symmetric spread of fresh water away from the well shown assumes a fairly homogeneous, isotropic aquifer with negligible regional flow relative to the flow rates induced by pumping during recharge or recovery. The actual configuration of the storage bubble may be considerably more complex. Modified from http://www.sfwmd.gov/org/pld/proj/asr/asrdefine.htm.

Concerns expressed about large-scale ASR in South Florida

ASR technology has been employed successfully in Florida since 1983 (Pyne, 1995), with individual well clusters having capacities up to about ten million gpd (38,000 m^3/day). However, the proposed scale in the CERP of 1.7 billion gpd (6.3 million m^3/day) is much larger than past projects. Implementation of ASR at the scale proposed in the CERP has raised a number of concerns among groundwater engineers and scientists in South Florida. Many of these concerns were outlined in a report prepared by the Aquifer Storage and Recovery Issue Team of the South Florida Ecosystem Restoration Working Group (ASR Issue Team, 1999) and presented to the Working Group in January 1999. The concerns addressed by the Issue Team, some of which were also noted in General Accounting Office (2000), were summarized in the following seven questions:

1. Are the proposed ASR source waters of suitable quality for recharge without extensive pretreatment?

2. What regional hydrogeologic information on the UFA is needed but unavailable for regional assessment?

3. Will the proposed ASR recharge volumes result in head increases sufficient to cause rock fracturing?

4. What will be the combined regional head increases from the regional scale ASR, and how will this affect individual ASR operation, change patterns of groundwater movement, and impact existing ASR wells, supply wells, or underground injection control (UIC) monitoring wells?

5. What are the likely water quality changes to the injected water resulting from movement and storage in the aquifer, and will the quality of the recovered water pose environmental or health concerns?

6. What, if any, is the potential impact of recovered water on mercury bioaccumulation in the surface environment?

7. What are the relationships among ASR storage zone properties, recovery rates, and recharge volumes?

These questions and others were viewed by the Committee on Restoration of the Greater Everglades Ecosystem (CROGEE) as falling into three major categories, which were ultimately reflected in the organization of the workshop and this report: regional science issues, water quality issues, and local performance/feasibility issues. These are reflected as chapters two, three, and four of this report.

The Lake Okeechobee and Western Hillsboro ASR Projects

This section contains a brief description of the overall plans for ASR in these two regions (Figure 2), including the pilot projects. It is taken almost entirely from USACE (1999). The project description, project background, and work breakdown structure from the Project Management Plans for the pilot projects are reproduced in Appendices E and F. The most recent drafts of the Project Management Plans, which are slightly revised from the drafts that the CROGEE evaluated, may be accessed at http://www.evergladesplan.org/projects/pilot/lake_o/lake_o_pp_main.htm and http://www.evergladesplan.org/projects/pilot/hillsboro/hillsboro_pp.htm, respectively.

Lake Okeechobee. This feature of the CERP includes ASR wells with a combined capacity of one billion gpd (3.8 million m^3/day) near Lake Okeechobee. These wells will be used in conjunction with a modified regulation schedule for the lake to achieve multiple-use purposes including storage capacity, flood control, and environmentally acceptable seasonal lake level fluctuations. About 200 wells with an assumed capacity of five million gpd (19,000 m^3/day) each are anticipated in the CERP. Some level of pre- and post-treatment of water recharged and recovered during ASR operation is also anticipated. Testing of treatment technologies, after characterization of source water, is one component of proposed pilot tests. Lake Okeechobee water will be injected into the UFA when lake levels are forecast to rise to undesirable levels. During dry periods, when lake levels are forecast to fall to undesirable levels, water would be recovered from wells and returned to the lake. It is assumed (based on existing ASR facilities) that recovery of aquifer-stored water will have no adverse effects on water quality in Lake Okeechobee, and may, in fact, improve it with respect to nutrient load. The pilot project will investigate changes to water chemistry resulting from aquifer storage and identify any necessary post-recovery treatment requirements to improve water quality.

The wells are designed to (1) provide additional regional storage with reduced evaporative losses and minimal land purchases compared to surface storage; (2) increase Lake Okeechobee's water storage capability to better meet regional water demands; (3) manage part of the regulatory releases from the lake primarily to improve Everglades hydropatterns and to meet supplemental regional water-supply demands of the Lower East Coast; (4) reduce harmful discharges to the St. Lucie and Caloosahatchee Estuaries; and (5) maintain and enhance current levels of flood protection.

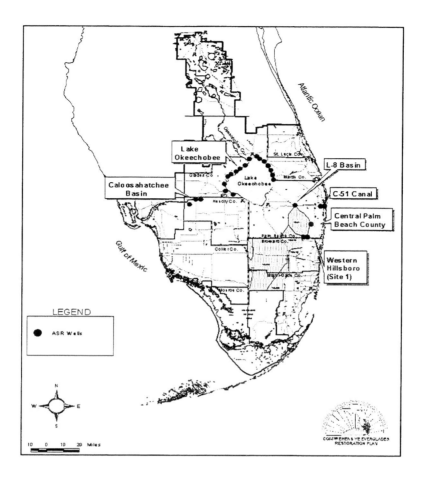

FIGURE 2. Map of the locations of the Lake Okeechobee, Western Hillsboro, and other planned ASR sites. Wells shown are for schematic purposes only; the actual total number of planned wells is about 333. From SFWMD (2000).

The pilot project is designed to provide benefit to environmental, urban and agricultural users. Its strategy is to install several exploratory/test ASR systems in geographically dispersed areas around the lake and to assess the feasibility of implementing ASR technology at a regional scale. Additionally, it will determine the water quality characteristics of waters to be injected, water recovered from the aquifer, and the water in the receiving aquifer. Information from the pilot project will provide the hydrogeological and geotechnical characteristics of the UFA at the pilot sites as well as a demonstration of the ability of the aquifer system to maintain injected water for future recovery.

Western Hillsboro. A series of ASR wells with associated pre- and post- water quality treatment will be located next to an aboveground reservoir with a storage capacity of about 15,000 acre-feet (4.89 billion gallons; 18.5 million m^3), or along the Hillsboro Canal. The combined ASR capacity would be about 150 million gpd (570,000 m^3/day). The initial design of the facility assumed 30 well clusters at 5 million gpd (19,000 m^3/day) per well. Some level of pre- and post-treatment of water recharged and recovered during ASR operation was anticipated in the CERP. Testing of treatment technologies, after characterization of source water, is one component of proposed pilot tests. Surface water and/or groundwater from the surficial aquifer adjacent to the reservoir would be the source of recharge water. The location, extent of treatment and number of ASR wells may be modified based on the results of the pilot project.

The purpose of this feature is to supplement water deliveries to the Hillsboro Canal during dry periods thereby reducing demands on Lake Okeechobee and the Loxahatchee National Wildlife Refuge. Surface water, possibly from the Hillsboro Canal, and shallow groundwater from along the margins of and below the reservoir, would be recharged when excess water is available, and would be released back to the canal to help maintain canal stages during the times of deficit.

The stated purposes of the pilot project are (1) to determine the most suitable sites and the optimum configuration for the ASR wells in the vicinity of the reservoir, (2) to evaluate many of the hydrogeological and geotechnical characteristics of the soils and aquifer in the area, and (3) to determine the specific water quality characteristics of water within the aquifer, water proposed for recharge, and water recovered from the aquifer.

Other ASR projects. The CERP also proposes four other sites for ASR (Figure 2). These include the C-43 Basin Storage Reservoir and Aquifer Storage and Recovery project in the Caloosahatchee River Region (44 wells), and three sites in Palm Beach County (total of 59 wells). All of these except the C-51 canal site (Figure 2) are associated with above-ground or in-ground reservoirs.

The Committee's Charge

The CROGEE's statement of task defines its role as "...provid[ing] scientific guidance to multiple agencies charged with restoration and preservation of... the greater Everglades.... In addition to strategic assessments and guidance, the NRC will provide more focused advice on technical topics of importance to the restoration efforts when appropriate." The CROGEE task of understanding and analyzing the ASR pilot projects was agreed upon as a high priority by both the Task Force and the NRC.

Before the first drafts of the Project Management Plans were written, the project design team had already received input from the ASR Issue Team, and had conducted other workshops around Florida. Between the first and second drafts, they received further input from various governmental agency staff. In the week before the workshop, the CROGEE received word that the USACE and SFWMD were proposing to request funds to conduct a regional study in parallel with the pilot project (Appendix C, topic 1). This was in response to feedback that the USACE and SFWMD had received from various individuals and institutions concerning the importance of understanding the regional hydrostratigraphic framework, aquifer properties, and hydrodynamics to evaluating the likely success of large-scale ASR.

The Project Management Plans have evolved further since the CROGEE held its workshop. According to Kwiatkowski (2000b), the major addition to the Lake Okeechobee ASR PMP in version 3 is the inclusion of three test/monitoring wells at the sites proposed for the exploratory/ASR wells. Major deletions from both PMPs in version three are tasks such as regional groundwater and geochemical modeling, additional monitoring wells and hydrogeologic investigations, regional source-water quality characterization study, and regional fracture analysis. Many of these tasks will be considered for inclusion in the proposed ASR Regional Study. It should be noted that as of this writing the proposed ASR Regional Study does not yet have a formal commitment of funding, nor has a budget been proposed.

Because of the fluid nature of what will or will not be included in the pilot projects (sensu stricto) and the regional ASR study, we consider this report to be a critique of the pilot projects *and related plans*. Unless otherwise specified, the term "feasibility studies" is used in this report to refer to the overall set of plans to answer key questions about local, regional and water quality issues with respect to the potential for success of ASR. It should be noted that this report is not designed to make judgements regarding the overall desirability of ASR as a major component of the CERP, either in absolute terms or in comparison with other storage options such as surface reservoir storage.

Finally, the entire CERP is based on the principle of adaptive assessment. Thus, the committee viewed this set of plans from the perspective of the extent to which they will contribute to process understanding that can improve design and implementation of restoration project components. In the next three chapters, we evaluate the feasibility studies from the perspective of whether they will address the major questions in the areas of regional science, water quality, and local performance/feasibility.

2

Regional Science Issues

Motivation

The Upper Floridan aquifer (UFA) extends across all of southern Florida and it is likely that ASR can be accomplished in this aquifer at some scale nearly anywhere in South Florida. However, assessing the impacts of ASR at the scale proposed in the CERP – 1.7 billion gpd (6.3 million m^3/day) – requires a regional, fairly detailed characterization of the hydrogeologic framework. The hydrogeologic framework information would form the basis for development of a numerical model capable of simulating the cumulative hydraulic effects of the proposed total number of ASR wells and their combined recharge and withdrawal rates. Without such a numerical simulation, the aggregate hydraulic impact of ASR wells on the existing UFA flow system cannot be quantified, and the feasibility of ASR at the scale proposed cannot be assessed reliably. The proposed ASR scale of 1.7 billion gpd (6.3 million m^3/day) is to be concentrated in a small area compared with the current total withdrawal of about 3 billion gpd (11 million m^3/day) that is pumped from the UFA throughout its areal extent of all of Florida, and parts of Alabama, Georgia, and South Carolina (Johnston and Bush, 1988). The project delivery team should be commended for recognizing the need for a regional study and for planning to pursue funding beyond the constraints of the currently funded pilot projects.

Issues Discussed

Workshop discussions included the items briefly listed below and expanded upon in the paragraphs that follow.

1. Identification of the information needed for adequately characterizing the hydrogeologic framework.
2. Compilation of existing information from reports on the hydrogeology of the UFA and from file data on existing wells and boreholes.
3. Assessment of how well the identified information needs are met by existing data and development of a test drilling program to fill in data gaps.
4. Coordination of surface geophysical surveys to augment drill-hole data.
5. Development of a conceptual model of the hydrogeology as a basis for construction of a numerical flow model.
6. Construction of an appropriate numerical flow model.
7. Application of the numerical model to predict head changes and provide the basis for assessing overall feasibility of proposed ASR scale and well deployment.

Conclusions and Recommendations

Based on the discussions at the workshop and also taking into account written material submitted prior to and after the workshop by the Project Development Team and others, the Committee on Restoration of the Greater Everglades Ecosystem (CROGEE) formulated the following recommendations.

Development of a preliminary list of information needs and compilation of available data should be undertaken as soon as possible. This exercise can be done for relatively low cost and within a short time can reveal where existing data are insufficient to meet identified needs. Categories of information needs include, but are not limited to, vertical distribution of potential recharge zones and their lateral extent, hydraulic properties, ambient water quality, and degree of confinement. Information on location, depth, and use of existing wells completed in the Floridan aquifer system is needed to assess potential impacts of proposed ASR wells on existing uses. Most of the existing well information is clustered in population centers along the coast. Little information is thought to be available inland where most of the ASR installations are proposed to be located.

A regional-scale three-dimensional numerical flow model will be required to assess impacts of proposed recharge and recovery volumes and rates on the UFA and adjacent hydrostratigraphic units. The model may need to account for variable density flow since the lower part of the UFA contains water that is somewhat saline, with several thousand or more mg/l of dissolved solids. Existing model codes based on an "equivalent porous medium" approach, such as HST3D (Kipp, 1997), are appropriate at a regional scale for estimation of head changes (Anderson and Woessner, 1992). Telescoped models of sub-regions, extracted from the regional model, also may be useful in assessing pressure buildup or drawdown at an intermediate scale. This regional and sub-regional modeling effort should be considered distinct from more detailed modeling that may be required to assess local scale feasibility questions related to recharge "bubble" growth and migration. Because of the solution conduit and fracture flow nature of the UFA, conventional models that employ an equivalent porous medium approach are likely to be poor simulators of solute transport in the vicinity of recharge wells (Long and Billeaux, 1987; Cacas et al., 1990). Development of suitable models for bubble growth and migration is considered in Chapter 4 of this report.

Development of the regional scale numerical flow model should proceed in parallel with initial conceptualization of the hydrogeologic framework. Sensitivity studies employing preliminary versions of the model can be used to identify data needs and gaps and thus help guide the planning of the test-drilling program. The regional study should have a budget adequate for the necessary drilling, core sampling, downhole geophysical logging, hydraulic testing and water quality sampling of these test wells. The tests and data collection should include all items planned for the exploratory ASR wells as specified in the Project Management Plan for the Lake Okeechobee Aquifer Storage and Recovery Pilot Project. The CROGEE concurs with existing plans to instrument observation wells with pressure recorders early in the regional study to obtain data with which to calibrate the numerical model.

Surface geophysical techniques, such as seismic reflection surveys, should be a component of the regional study. Such surveys can help to constrain the three-dimensional geometry and continuity of hydrostratigraphic units (Kindinger et al., 1999), especially when used in conjunction with results from a relatively limited number of new exploratory wells. The surveys would provide a means for extrapolating detailed information collected at test well sites across broad areas where well control is sparse, thereby improving the quality of input to the regional-scale flow model. Surveys that use land-based techniques, as well as those that employ marine techniques (across Lake Okeechobee or along canals), should be considered. Without the geophysical surveys, a much more extensive and more costly drilling program may be required to adequately characterize the hydrogeology.

Once constructed, the regional flow model should be used to assess potential impacts related to full-scale implementation of ASR. Aggregate head buildup or drawdown estimated from the model can be used to predict changes in direction and velocity of flow in the UFA, and to evaluate the possible consequences of these changes on existing wells (see Bradbury and Krohelski, 1999 for a similar approach). The aggregate head buildup is needed also to predict areas where more detailed studies of the potential for hydraulic fracturing (see chapter 4) should be done. An elevated potentiometric surface that could induce fracturing and upward migration of recharged water could result from the superposition of head buildups of multiple ASR wells.

The regional or telescoped models of sub-regions can be used to compare and evaluate alternative ASR system designs. For example, regional effects of an ASR system that is concentrated in a small geographical area, such as the planned deployment of 100 ASR wells along the northern half of Lake Okeechobee, could be compared to a system of similar volume that is more geographically dispersed. Various recharge and withdrawal scenarios can be easily simulated with the regional model, and the model results can be used to optimize the ASR system while minimizing deleterious impacts (e.g., using the approach described by Wagner, 1995).

The regional study should include development of a formal procedure for the siting of ASR wells. For the few exploratory ASR wells included in the pilot project, preliminary siting was based solely on location of a water source for recharge and proximity of a water body for receipt and conveyance of drilling and testing fluids. However, for full implementation of the ASR proposal, well siting should include evaluation of other factors (e.g., anticipated well capacity, proximity to other users, etc.) as well. It is important that the site selection process be explicit, repeatable, and based on best available data and selection methods.

The proposed Phase 1 siting analyses include simple overlay of GIS data to evaluate site suitability. This approach may be adequate for the local pilots, but a more formal, multi-criteria siting approach should be designed and tested during the pilot studies that would facilitate the much larger and more complex problem of siting many wells over the region. Promising approaches range from structured scoring methods such as Analytical Hierarchy Process and Knowledge Bases (Reynolds et al., 1999) to more complex optimization models using mathematical programming or heuristic search procedures (e.g. Haight et al., 2000; Matthews et al., 1998; Miller, 1996; Murray et al., 1998).

In conclusion, the pilot projects provide a valuable means for acquiring detailed information on ASR performance at a few specific sites. However, even if all the sites tested prove successful, they will not by themselves demonstrate the feasibility of ASR implementation regionally at the scale of 1.7 billion gpd (6.3 million m^3/day). A regional study involving construction of a regional flow model is an invaluable and indispensable tool to assess feasibility of ASR at the proposed scale.

3

Water Quality Issues

Motivation

Although aquifer storage and recovery (ASR) has been used as a means of supplementing water supplies in Florida for over a decade, the unprecedented scale of the ASR program proposed in the CERP raises a wide range of water quality issues that go beyond the concerns associated with local ASR projects. In part, these issues arise because of the chemical diversity of the water bodies and aquifers involved in ASR across the large geographic range of the CERP. The issues also arise because ASR water may be used for agriculture, for augmenting water inflows to natural ecosystems in the Everglades, and, indirectly, to supplement municipal (drinking water) supplies. Different regulations and different concerns about water quality arise in connection with these different intended uses.

In practical terms, answers to a range of water quality questions are critical to the regulatory process for permitting the ASR wells and the release of recovered water to Lake Okeechobee and various canals. The ASR pilot program thus has placed an appropriately heavy emphasis on collecting an extensive array of water quality measurements. However, chemical analyses on water recharged to and recovered from pilot wells cannot answer many questions about potential biological impacts of ASR water on the Everglades ecosystem. Such analyses by themselves also cannot provide the mechanistic information needed to develop geochemical models to predict how water chemistry/water quality will change in a full-scale ASR program. Consequently, the Committee on Restoration of the Greater Everglades Ecosystem (CROGEE) concludes that the water quality measurements proposed for the ASR pilot program should be supplemented by laboratory and field-scale investigations.

Issues Discussed

Workshop discussions related to water quality issues included the following topics. The list is not intended to suggest an order of importance, nor does it reflect the amounts of time spent discussing topics during the formal session on water quality. Some topics on the list represent an integration of several related themes or issues raised during the workshop.

1. Adequacy of existing regulatory standards for water quality to assure that the extensive use of ASR in the Everglades restoration program has minimal negative impacts on ecosystems.

2. The need for ecotoxicological studies and bioassays at laboratory and field scales to elucidate potential impacts of water quality changes including not only concentrations of nutrients and trace metals, but also concentrations of sulfate and chloride and properties such as pH and alkalinity.

3. Need for and nature of studies on microbial (pathogen) survival to be conducted during the pilot projects. Particular concern was expressed about use of coliform bacteria to represent the pathogens present in surface water sources used for aquifer recharge.

4. Effects of dissolved and particulate organic matter in recharge water on biogeochemical processes in the subsurface environment and the potentially adverse effects of these processes on ASR system functioning.

5. Rates of mineral dissolution reactions in the aquifer storage zone that could release heavy metals, arsenic, radionuclides, and major ions from the geological matrix into water that is to be recovered for use as drinking water or to supplement flows to the Everglades.

6. Mechanisms of mixing of relatively dilute recharge water with more saline pore fluids in the storage zone, which will affect the extent to which water quality is changed during the aquifer storage and recovery process.

Conclusions and Recommendations

The CROGEE concludes that ASR water used in the Everglades restoration will probably need to achieve concentrations for some variables that are lower than existing Florida numerical water quality criteria for drinking water. Likewise, Florida's existing Class III water quality criteria (i.e., criteria for protection of fish and wildlife) may not be sufficient to ensure that use of ASR water does not have negative ecological impacts. This is partly due to the great diversity in chemical composition among surface waters within South Florida. For example, surface waters in the southern part of the Everglades have low concentrations of ions, whereas waters in Lake Okeechobee and waters just to the south of the lake have relatively high ionic content. The species composition of plant life growing in these waters reflects the chemical composition of the water. Specifically, native flora of the soft water in the southern Everglades are adapted to those conditions. If the use of recovered ASR water during periods of low rainfall increases the ionic strength of Everglades water (see Mirecki et al., 1998, for example), the composition of the plant community may change significantly even though the water may meet all existing criteria.

The likelihood of such changes cannot be determined merely by conducting chemical analyses as part of the pilot program nor even by short-term laboratory-scale bioassays. The CROGEE recommends that ecotoxicological studies, including long-term bioassays conducted at the field scale, be undertaken during the period of the pilot program to evaluate the ecological impacts of water quality changes resulting from broad use of ASR water to supplement inflows to the Everglades during droughts.

Water recharged to ASR wells will come primarily from Lake Okeechobee and other impoundments. The source waters and surrounding lands can and should be managed to minimize the occurrence of pathogens in the water being pumped down the wells. However, because these surface water bodies may contain viruses and bacteria that are not naturally present in the aquifer system, the CROGEE recognizes that pathogen die-off studies need to be performed as part of the permitting process. It recommends that those studies be done under conditions as realistic of the full-scale ASR process as possible. For example, aquifer materials and their resident microbial communities can be returned to the laboratory for experimentation. In instances in which it is impossible to obtain core segments, drill cuttings can be substituted, although this may result in a substantial loss in the quantitative value of the results. In particular, die-off studies should focus on the pathogenic species of concern, especially viruses, rather than on model microorganisms, whose survival rates may not necessarily mimic those of actual pathogenic species.

The source water for recharge will be relatively high in dissolved organic matter, and, depending on the degree of filtration, may contain some particulate carbon as well. The biodegradation of this material within the aquifer could have major impacts on aquifer chemistry in the recharge zone. Increased oxygen demand imposed by this organic matter could stimulate nitrate and sulfate reduction, generating low redox potentials and high sulfide levels. In turn, this could stimulate methylation of mercury and reductive dissolution of mineral phases, possibly increasing the dissolved concentrations of heavy metals, arsenic and radionuclides in the water being stored for recovery. Although the alkalinity of the source water is likely to be moderately high, and the waters thus should be well-buffered with respect to pH, production of carbon dioxide from the biodegradation of organic matter could lower the pH of the stored water. This could affect a variety of chemical processes and promote heavy metal dissolution. Microbial growth stimulated by the dissolved organic matter could form biofilms on mineral phases in the aquifer. While this process could protect some phases from dissolution, biofilms could possibly accelerate other dissolution reactions. Thus, the characterization of organic carbon in the source water should be a priority, as should studies designed to anticipate the effects of this material on biogeochemical processes in the subsurface.

In addition, the effects of chlorination as a proposed pre-treatment process for recharge water in the Western Hillsboro ASR need to be evaluated carefully with respect to potential toxic halogenated organic compounds (e.g., Thomas et al., 2000; Landmeyer et al., 2000). The usefulness of other treatment options, such as ultraviolet radiation, should be studied during the pilot project.

The potential effects of these changes on the chemical quality of recovered water need to be examined during the pilot study. While routine water quality monitoring conducted as part of the pilot well studies will provide important information, such studies are not sufficient to answer questions about the mechanisms that cause observed changes in water quality. Consequently, monitoring alone will not provide the required basis to develop predictive models of important redox and dissolution/precipitation reactions under full-scale, long-term implementation of ASR. The CROGEE thus recommends that additional laboratory experiments and chemical modeling be undertaken during the pilot phase to address these issues in a scientifically defensible way. Process studies can be conducted under controlled conditions if aquifer materials and their resident microbial communities are returned to the laboratory for experimentation. Incremental core segments can be used to examine the rates of dissolution and precipitation, biological oxidation and reduction reactions and the movement and reaction of the products to determine both rates and extent of kinetically controlled phenomena. As with pathogen studies, core segments are superior to drill cuttings for this work, because the drilling process destroys the rock fabric and creates uncertainty with respect to the depth of origin of a given cutting.

Finally, a better understanding of the mechanisms responsible for mixing relatively dilute recharged water with more saline pore fluids in the storage zone is essential for anticipating changes in dissolved solids during ASR. The transport and mixing processes in the Upper Floridian aquifer may not conform to the "equivalent porous medium" model commonly used in groundwater flow models. Elucidating these processes should be an objective of the pilot studies (see chapter 4).

4

Local Performance/Feasibility Issues

Motivation

The feasibility of injecting, storing, and recovering specified volumes of water at individual ASR wells and in local clusters of ASR wells is the issue to which the authorized pilot projects at Lake Okeechobee and Western Hillsboro are most specifically directed. The draft project management plans that were available for review prior to the Committee on Restoration of the Greater Everglades Ecosystem (CROGEE) workshop indicated that the general locations of exploratory wells that will be drilled as part of these pilot studies have already been selected. These exploratory wells ultimately will be converted to pilot ASR wells for cycle testing. As described in the draft project management plans, the tests at these wells can be expected to provide reliable information on feasibility at the specific locations selected for testing. However, in keeping with the principle of adaptive assessment, the CROGEE judges that the pilot projects should not be considered simply as demonstrations at particular sites. Instead, they should be viewed as an opportunity to develop a better understanding of the hydrogeologic and well construction characteristics that control the relationships between storage intervals, recharge volumes, and recoverability.

The discussion during this period of the workshop was intended to highlight design considerations for the pilot projects that could maximize the value of the data obtained for improving understanding of ASR performance at both the specific sites tested and more generally in the Upper Floridan aquifer (UFA). The better understanding that would result from carefully designed pilot studies then could be translated into improved siting and design of additional ASR wells both in the immediate vicinity of the pilot studies and elsewhere in the UFA.

Issues Discussed

Workshop discussions included questions, comments and responses related to the items listed below.

1. Selection criteria used to determine locations for proposed exploratory wells.
2. Selection criteria to be used for selection of target recharge intervals.
3. Advantages and disadvantages of using short or long open intervals in ASR wells. Short intervals may simplify aquifer management and minimize in-well circulation, but may limit recharge rates.
4. Design of monitoring wells to allow monitoring of "bubble" evolution, to evaluate potential leakage across confining units, and to assess well interference effects for pilot clusters.
5. Availability of models to simulate site-scale processes for comparison with monitoring data collected during cycle testing.
6. Considerations of trade-offs between rates of recharge and recoverability.

7. Quantification of efficiency of recovery, recoverability, or other measures of "success," particularly when considering water quality standards of the surface water bodies receiving recovered water.

8. Time scales of cycle testing and how these relate to anticipated operational cycles in recharge and recovery.

9. Use of data that may be available from studies of existing ASR systems to address questions related to feasibility.

10. Whether the pilot studies alone or in conjunction with other studies will provide sufficient information to demonstrate feasibility or infeasibility of a regional scale ASR system.

Conclusions and Recommendations

Based on the discussions at the workshop and also considering written material submitted prior to and after the workshop by the Project Development Team and others, the CROGEE formulated the following recommendations.

Test conditions

In order to maximize transferability of pilot project results to design and implementation of regional scale ASR systems, the pilot projects should be designed as tests to elucidate processes of "bubble" formation and migration under a variety of conditions. Some of these conditions include long and short open intervals of recharge wells, high and low injection pressures, homogeneous and heterogeneous aquifer properties within the storage zone, presence or absence of highly transmissive beds within the storage zone, greater and lesser contrasts in salinity between injected and native water, and long and short periods of storage prior to recovery.

Tests designed to compare effects of short and long open borehole sections in the ASR wells would be particularly useful to improving understanding of factors important to ASR well design at the selected pilot facilities and, more generally, throughout the UFA and in other aquifers with significant vertical variations in hydraulic properties. One possible way to conduct such tests within the constraints of a limited total number of pilot wells would be to construct an ASR well with multiple open sections for testing individually or in combination.

Monitoring wells

An important component of any test design is the monitoring strategy. Adequate monitoring is essential to obtaining data that can be used to develop conceptual models of bubble development and to test and calibrate numerical models of the relevant processes. Multiple monitoring wells at each pilot ASR site should be a high priority to delineate the geometry and heterogeneity of fresh water migration during recharge. The shape and heterogeneity of the fresh water bubble may ultimately dictate the extent of mixing between injected and ambient pore water during storage. This mixing is one of the important factors in limiting efficiency or recoverability. Multiple monitoring wells also are required to obtain better estimates of magnitudes, spatial variability and possible anisotropy of hydraulic properties of aquifers and confining units. Nested monitoring wells, or monitoring wells that can be packed off and sampled from discrete intervals, should be constructed for purposes of delineating potential preferential flow paths from the ASR wells and for quantifying mixing between relatively dilute recharged water and more saline ambient pore water in the storage zone.

The draft project management plans suggest that only a single UFA monitoring well and a single surficial aquifer well are anticipated in the conceptual designs for ASR pilot facilities. The CROGEE recognizes that significant costs would be associated with additional monitoring beyond those included in the initial plans. In case of budget constraints, the CROGEE concludes that it would be preferable to do more detailed studies, with enhanced monitoring, at a reduced number of pilot ASR sites or with a reduced number of exploratory ASR wells, rather than to do more limited studies using the currently planned number of exploratory wells but with too few monitoring wells.

Cycle duration

During the workshop discussion it was noted that data are available from a number of existing ASR systems that have operated for over a decade. These existing data can be quite useful in design of the pilot projects and in anticipating the hydraulic performance of ASR wells in general. However, because existing ASR systems that are used for water supply management generally operate on relatively short cycle periods, it may be difficult to use results from those systems to predict the degree of mixing and water-rock interactions during longer-term recharge and storage. Such mixing and water quality changes due to water-rock interactions may be the limiting factors in efficiency or recoverability of the regional scale ASRs proposed as part of the CERP. Design of cycle testing of the pilot wells must take into account the anticipated duration of recharge and recovery cycles that will be employed during the final operation. The hydrologic models used for design of the CERP indicate that continuous recharge is likely to be required in the Lake Okeechobee regional ASR for multi-year periods. This suggests that continued recharge for periods on the order of a year or more prior to recovery should be employed for the Lake Okeechobee pilot ASR cycle testing in order to obtain data that will be relevant to operational performance.

Interpretation of results

Observations obtained from a suitable network of monitoring wells can provide the basis for development of new or improved conceptual and numerical models to simulate "bubble" development and migration. Additional data may also be available from monitoring of other existing ASR systems, and this information would also be useful in developing and testing such models. Numerical codes used to test conceptual models and to simulate bubble migration may need to account for three-dimensional, density dependent flow and solute transport in multi-permeability media containing fractures and solution conduits. The CROGEE is not aware of any existing codes that have the full capability to accommodate these aquifer characteristics. Thus, construction of models for these local scale processes is likely to require substantial technical effort with a commensurate level of funding.

Additional considerations

While the pilot projects are likely to be most useful in assessing local scale feasibility issues, they can also contribute data to the assessment of the regional hydrogeologic framework and aquifer characteristics. Use of data collected from the exploratory wells as part of a regional scale aquifer characterization is anticipated in the draft project management plans that were available prior to the workshop. The CROGEE encourages such use of data collected at the pilot sites, noting that careful coordination between the pilot studies and the regional study will maximize the value of these data to understanding the system at a variety of scales.

Fracture potential studies, which may be limited to desk-top analyses or expanded to more rigorous modeling approaches, are described briefly in the draft project management plans. The CROGEE recognizes the importance of such studies and commends the project management team for including them in the plans. The pilot studies should provide valuable additional information on the potential for fracture development due to a single ASR well or cluster. Results of fracture potential studies conducted as part of the pilot projects should be combined with results of the regional hydrogeologic models in order to assess the potential for fracturing during full-scale ASR operation.

5
Summary and Conclusions

The Committee on Restoration of the Greater Everglades Ecosystem (CROGEE) has identified three general areas in which additional data and studies are required to address uncertainties related to feasibility and optimal design of regional ASR systems intended to provide storage required by the CERP. First, an improved understanding of the regional hydrogeologic framework, at the level of detail required to construct a regional scale numerical model of groundwater flow, is essential to assessing regional scale feasibility and impacts of the proposed regional ASR systems. Second, an improved understanding of water quality changes that will occur during storage of freshwater in the subsurface, and of the effects of modified water quality on human and ecological recipients of this water, is required to determine the chemical characteristics that can be used in assessments of recovery efficiency or recoverability at individual ASR wells. Third, an improved understanding of how local hydrogeologic properties (e.g., vertical distribution of hydraulic conductivity) and well construction features (e.g., location and length of open interval for recharge and recovery) interact to enhance or limit well capacity and recovery efficiency is required for optimal design of individual wells and well clusters.

Better understanding in these three areas of uncertainty will require studies that extend beyond the scope of the ASR pilot projects, at least as they have been described in draft project management plans. In these plans, it appears that the pilot studies are almost exclusively devoted to local feasibility issues. A regional study and modeling effort has been proposed recently as a separate project, but it has not been authorized or funded. While the pilot projects will generate some of the data needed to anticipate water quality changes during storage, studies to assess the appropriate water quality standards for discharge of recovered water to ecosystems are not part of any current or planned projects of the CERP. The pilot projects could, in principle, provide data needed to elucidate relationships between hydrogeologic properties and well construction characteristics. However, the pilot studies will be successful in providing such data only if they are carefully designed to test and monitor a variety of storage intervals under a range of well construction and operational conditions.

Many of the recommendations in the preceding chapters have been made previously by other groups, in particular the ASR Issue Team. Foremost among these is the need for a regional hydrogeologic synthesis and model to allow assessment of regional-scale feasibility and impacts of the proposed regional ASRs. Ideally the regional synthesis and modeling would precede any local scale feasibility studies such as those planned as part of the authorized pilot studies. This would allow the site selection for pilot studies to make use of regional information and also could provide a mechanism for using the pilot study sites to fill critical data gaps in the regional hydrogeologic framework. Given that the planning for the pilot studies already is well underway, it is even more important that a regional study be funded and initiated in the near future to optimize coordination of data collection at local and regional scales.

Information from the regional study, water quality studies, and local scale pilot studies ultimately must be synthesized and used together in the overall assessment of ASR as a component of the CERP. With an improved understanding of ASR systems that will result from these studies, it should be possible to develop ASR system designs that then can be compared with other storage options in terms of overall performance, including storage capacity and cost effectiveness. This will require examination of related issues that were not addressed in any detail during the workshop, but are important to the overall assessment of ASR feasibility and effectiveness. Some of the topics that should be addressed in the design and evaluation of ASR systems are listed below.

- Once information from this work is available, it may be important to refine the analysis of overall performance of surface and subsurface storage, particularly if anticipated recovery efficiency for ASR differs significantly from the value of 70% assumed in model runs used to develop the CERP.
- Estimated evaporation losses from surface reservoirs should be compared with anticipated subsurface losses to the aquifer during ASR in order to quantify relative performance of surface and subsurface storage options.
- A significant increase or decrease in recovery efficiency would dictate re-examination of the number of wells required for a regional ASR system.
- Similarly, if recharge capacities for efficient operation of ASR wells differ significantly from 5 million gpd (19,000 m^3/day) assumed in the CERP, it may also be necessary to re-evaluate the number of wells required to provide the necessary storage capacity.
- Estimates of energy costs for long-term operation should include contingencies for possible changes in fuel costs over the anticipated project life.

As a final comment, the CROGEE notes that the CERP calls for ASR to be implemented in phases. The Committee agrees that phased implementation is an appropriate strategy and strongly recommends a) thorough evaluation of the environmental effects of each incremental increase in scale of ASR, and b) ongoing adaptive assessment of the program.

References

Anderson, M.P. and Woessner, W.W. 1992. Applied groundwater modeling: simulation of flow and advective transport. San Diego: Academic Press, Inc.

Aquifer Storage and Recovery Issue Team. 1999. Assessment report and comprehensive strategy: a report to the South Florida Ecosystem Restoration Working Group. [Online]. Available: http://www.sfrestore.org/teams/asr/documents/asrreport.htm [2000, November 17].

Bradbury, K.R. and Krohelski, J.T. 1999. A regional groundwater model as a tool for evaluating the impacts of urbanization. Geol. Soc. of Amer. Abstr. with Prog. 31(7):A-156.

Cacas, M.C., Ledoux, E., de Marsily, G., Tillie, B., Barbreau, A., Durand, E., Feuga, B., and Peaudecerf, P. 1990. Modeling fracture flow with a stochastic discrete fracture network: calibration and validation. 2. The transport model. Water Resources Res. 26:491-500.

CH2M Hill. 1989. Construction and resting of the aquifer storage and recovery (ASR) demonstration project for Lake Okeechobee, Florida. Report for the South Florida Water Management District.

General Accounting Office. 2000. Comprehensive Everglades Restoration Plan: additional water quality projects may be needed and could increase costs. GAO/T-RCED-00-297. Washington, D.C.

Haight, R.G., Revelle, C.S., and Snyder, S.A. 2000. An integer optimization approach to a probabilistic reserve site selection problem. Operations Research 48: (5) 697-708.

Johnston, R.H. and Bush, P.W. 1988. Summary of the hydrology of the Floridan aquifer system in Florida and in parts of Georgia, South Carolina, and Alabama. U.S. Geological Survey Professional Paper 1403-A. Reston, Va.: U.S. Geological Survey.

Kindinger, J.L., Davis, J.B., and Flocks, J.G. 1999. Geology and evolution of lakes in north-central Florida. Environmental Geology 38:301-323.

Kipp, K.L., Jr. 1997. Guide to the revised heat and solute transport simulator, HST3D--Version 2: U.S. Geological Survey Water-Resources Investigations Report 97-4157. Reston, Va.: U.S. Geological Survey.

Kwiatkowski, P. 2000a. ASR pilot projects for the Comprehensive Everglades Restoration Plan (CERP). Presentation to the Committee on Restoration of the Greater Everglades Ecosystem (CROGEE). Miami, FL, October 19, 2000.

Kwiatkowski, P. 2000b. Draft meeting summary, Project Delivery Team meeting for Lake Okeechobee and Western Hillsboro ASR pilot projects. West Palm Beach, FL, November 13, 2000.

Landmeyer J.E., Bradley, P.M., and Thomas, J.M. 2000. Biodegradation of disinfection byproducts as a potential removal process during aquifer storage recovery. Journal of the American Water Resources Association 36: 861-867Long, J.C. and Billeaux, D.M. 1987. From field data to fracture network modeling: an example incorporating spatial structure. Water Resources Res. 23: 1201-1216.

Matthews, K.B., Sibbald, A.R. and Craw, S. 1999. Implementation of a spatial decision support system for rural land use planning: integrating geographic information system and environmental models with search and optimisation algorithms. Computers and Electronics in Agriculture 23: 9-26.

Miller, H.J. 1996. GIS and geometric representation in facility location problems. International Journal of Geographic Information Systems 10: (7) 791-816.

Mirecki, J.E., Campbell, B.G., Conlon, K.J., and Petkewich, M.D. 1998. Solute changes during aquifer storage recovery testing in a limestone clastic aquifer, Ground Water 36: 394-403.

Murray, A.T., Church, R.L., Gerrard, R.A., and Tsui, W.S. 1998. Impact models for siting undesirable facilities. Papers in Regional Science 77: 19-36.

National Research Council (NRC). 1994. Ground water recharge using waters of impaired quality. Washington, D.C.: National Academy Press.

Pyne, R.D.G. 1995. Groundwater recharge and wells: a guide to aquifer storage recovery. Boca Raton, Florida: Lewis Publishers.

Reynolds, K.M., Jensen, M., Andreasen, J. and Goodman, I. 1999. Knowledge-based assessment of watershed condition. Computers and Electronics in Agriculture 23: 315-334.

South Florida Water Management District (SFWMD). 2000. Everglades Consolidated Report. West Palm Beach, FL: SFWMD.

Thomas, J.M., McKay, W.A., Cole, E., Landmeyer, J.E., and Bradley, P.M. 2000. The fate of haloacetic acids and trihalomethanes in an aquifer storage and recovery program, Las Vegas, Nevada. Ground Water 38: 605-614.

United States Army Corps of Engineers (USACE). 1999. C&SF Restudy final integrated feasibility report and programmatic environmental impact statement (PEIS). Jacksonville, FL: USACE.

Wagner, B.J. 1995. Recent advances in simulation-optimization groundwater management modeling. Rev. Geophys. Supp., p. 1021-1028.

APPENDIXES

Appendix A

Workshop Agenda

Committee on Restoration of the Greater Everglades Ecosystem (CROGEE)

ASR Workshop
Wyndham Miami Biscayne Bay, Miami, Florida
October 19, 2000

Agenda

Thursday, October 19
(Mira Flores Meeting Room)

8:00 a.m.	Registration
8:20 a.m.	Introductions, charge of committee, format of workshop
8:30 a.m.	Summary of ASR pilot projects by project delivery team
9:00 a.m.	**Topic I - Regional science issues**
	9:00 Introduction
	9:10 Panel/committee questions
	9:55 Open discussion
	10:15 Subcommittee deliberations
10:25 a.m.	Break
10:40 a.m.	**Topic II - Water quality issues**
	10:40 Introduction
	10:50 Panel/committee questions
	11:35 Open discussion
	11:55 Subcommittee deliberations
12:05 p.m.	Lunch *(Serena Meeting Room)*
1:05 p.m.	**Topic III – Local performance/feasibility issues**
	1:05 Introduction
	1:15 Panel/committee questions
	2:00 Open discussion
	2:20 Subcommittee deliberations
2:30 p.m.	Break
2:45 p.m.	Other issues (from committee, SFWMD, audience), left-over issues, "open mike"
4:30 p.m.	Adjourn

NOTES FOR ASR WORKSHOP:

We interpret the essential purpose of the pilot projects as answering questions such as:

It is physically possible that the quantities of water envisioned by the CERP can be stored and recovered in the Florida aquifer at the proposed spatial and temporal scales?

If so, what undesirable physical, chemical, ecological, and societal impacts may ASR cause, and how do we assess the likelihood and seriousness of these impacts?

Two essential questions that will govern the execution of the workshop are:

As designed, will the pilot projects largely answer these questions?

If not, how might they be improved?

Appendix B
Workshop Participants

CROGEE Members

Jean Bahr
Patrick Brezonik
James Davidson
Wayne Huber
Pete Loucks
Kenneth Potter
Larry Robinson
John Vecchioli

Invited Experts

Joan Browder, Southeast Fisheries Science Center, NOAA
James Cowart, Florida State University
Rich Deuerling, Florida Department of Environmental Protection
Richard Harvey, U.S. Environmental Protection Agency
Donald McNeill, University of Miami
Tom Missimer, CDM-Missimer International
Mark Pearce, Water Resource Solutions Inc.
Robert Renken, U.S. Geological Survey
Joan Rose, University of South Florida
Walt Schmidt, Florida Geological Survey

Other Attendees

Nick Aumen, NPS/Everglades
Brad Brown, NOAA
Chris Brown, USACE
Edwin Brown, USACE
Jose, L. Calas, FDEP
Kevin Cunningham, US Geological Survey
Carol Daniels, NPS, Tallahassee
Bill Dobson, Miami-Dade County
Paul Dresler, DOI
Julio Fanjul, SFERTF
Juanita Greene, Friends of the Everglades
Lloyd Horvath, Water Resource Solutions
Aaron Higer, USGS
Muhammad Irfan, USACE Jacksonville District
Jack Kindinger, USGS
Shawn Komlos, Audubon of Florida
Pete Kwiatkowski, SFWMD
Glenn Landers, USACE Jacksonville District

Bill Neimes, USACE Jacksonville District
Rick Nevulis, SFWMD
Peter Ortner, NOAA
David Pyne, CH2M HILL
Ron Reese, USGS
Peter Rosendahl, Florida Crystals
Tom Teets, SFWMD
Bob Verrastro, SFWMD
Mike Waldon, USFWS
Dann Yobbi, USGS

NRC Staff

William Logan
Stephen Parker
David Policansky

Appendix C

Questions sent to the SFWMD prior to the workshop, and its responses

October 17, 2000

The following is the U.S. Army Corps of Engineers (USACE)/South Florida Water Management District (SFWMD) preliminary response to written questions posed by the CROGEE ASR Subcommittee. The response is formatted such that the question is repeated, followed by the USACE/SFWMD response in bold. We hope that providing these comments ahead of the October 19, 2000 meeting will clarify some issues and guide the discussions at the meeting.

Topic I - Regional hydrogeologic framework issues (Note that this is primarily portions of Issue Team items 2 and 4)

Questions that are likely to arise in some form:

- What are the main sources of uncertainty with respect to the regional hydrogeologic framework?

Perhaps the greatest source of uncertainty is the fact that the existing hydrogeologic data have not yet been compiled and evaluated. Regional hydrogeologic investigations conducted by the USGS as part of its Regional Aquifer System Analysis (RASA) program, as well as the Florida Geological Survey's (FGS's) evaluation of the Hawthorn Group (Scott, 1988) provide solid evidence as to the lateral extent and continuity of the Floridan Aquifer System (FAS) and the overlying Hawthorn confining unit.

Recognizing the importance of an understanding of regional issues, and the budget limitations of the pilot projects, USACE/SFWMD are now proposing to allocate funds to conduct a Regional Study in parallel with the pilot projects. The scope of the Regional Study is not yet defined, but will be refined based on CROGEE comments.

- Do the pilot projects include the necessary studies to resolve these uncertainties and fill in the data gaps to characterize the properties and regional extent of possible injection zones as well as properties and extent of other significant hydrostratigraphic units (e.g. aquifers that are used for drinking water supply, confining units that would separate these from the injection zones or separate ASR injection zones from intervals used for waste injection)?

Development of the regional hydrogeologic framework will be one of the first tasks conducted. Filling in data gaps with additional wells is a logical but costly task, and we believe beyond the scope of the pilot projects. If needed, additional wells could be installed as part of the recently proposed Regional Study.

Note that we propose to install three (3) monitor wells at geographically dispersed locations around Lake Okeechobee to get some site-specific data geared toward evaluation of the

Appendix C: Questions to the SFWMD and its Responses

potential for ASR zones and confining units. This task will occur simultaneously with development of the hydrogeologic literature search. Following this task, three (3) large-diameter exploratory wells will be permitted and installed to better assess hydrogeologic characteristics, evaluate hydraulic capacity, and result in selection of the target ASR storage zone.

- Will the pilot projects be adequate to develop a regional conceptual model suitable for translation into the numerical model that is intended to provide a preliminary assessment of regional effects of the ASR system?

Recognizing the importance of an understanding of regional issues, and the budget limitations of the pilot projects, USACE/SFWMD are now proposing to allocate funds to conduct a Regional Study in parallel with the pilot projects. The scope of the Regional Study is not yet defined, but will be refined based on CROGEE comments. The scope of the Regional Study will include development of a conceptual model from several sources including:

1. Hydrogeologic literature search
2. Automation of 30 FAS wells to assist in model calibration
3. Hydrogeologic data from the 3 new monitor wells noted above
4. Hydrogeologic data from the 3 new large-diameter exploratory wells

We believe these data are sufficient to develop a regional numerical model, but are aware that one can always make an argument that more data are needed.

- If the pilot projects will not provide all of the necessary information to characterize the regional framework and to allow development of a numerical model, what other studies will be going on in parallel to provide the necessary information?

Recognizing the importance of an understanding of regional issues, and the budget limitations of the pilot projects, USACE/SFWMD are now proposing to allocate funds to conduct a Regional Study in parallel with the pilot projects. The scope of the Regional Study is not yet defined, but will be refined based on CROGEE comments.

- What information is available, or what analysis has been done, to estimate changes to the regional flow system due to the aggregate head-buildup and drawdown associated with the proposed projects?

No analysis has been conducted to date. Anecdotal data from existing FAS and ASR Wellfields (Town of Jupiter, Cocoa, Peace River) indicate that intense pressure build-up and drawdown has not occurred. We will document these and other data during the plan formulation process. Ultimately, the proposed regional modeling effort will best allow us to address this question.

- How did selection of the proposed locations for exploratory wells, pilot ASR wells and monitor wells relate to locations that are likely to provide the necessary constraints for characterizing the regional framework?

The proposed sites for the 3 exploratory wells (i.e., at the confluence of the Caloosahatchee, Kissimmee, and St. Lucie Rivers with Lake Okeechobee) were selected for two primary reasons. First, these sites are known to possess a great amount of water availability. Of course, we know that Lake Okeechobee has available water, but conversations with Lake experts indicate that water quality can vary substantially between the Lake and rivers/tributaries, depending on seasonal and weather conditions. The intent here is to allow for ASR recharge with the best quality water when it is available, be it Lake Okeechobee or its tributaries/estuaries. Second, we must discharge aquifer water (e.g., reverse-air drilling, well development, packer test, specific capacity test, aquifer performance test, etc.) to facilitate well construction and testing. This must be accomplished in compliance with the required National Pollutant Discharge Elimination System (NPDES) permits. Often times, it is difficult to meet the NPDES permit requirements (e.g., 800-meter mixing zone for conductivity, no exceedance of 29 NTUs above background turbidity) unless a flowing water body is close by. Thus, the proposed monitor and exploratory wells are located at these major tributaries/estuaries.

These reasons for site selection, however, do not preclude the ability nor minimize the importance of gathering useful regional hydrogeologic information. In fact, we believe it prudent to begin the regional characterization at those sites where ASR systems are most likely to be situated.

- Could ASR exploratory wells, downsized, be used in a hydrostratigraphy mode and later converted to ASR monitor wells?

Yes, but it is our intent to re-permit these exploratory wells as functional ASR wells, depending on source-water quality.

- How will results of geophysical logging and hydraulic tests at the proposed exploratory well locations be extrapolated to provide information on hydraulic properties over the regional scale?

By comparing the regional hydrogeologic data with that from the monitor wells and exploratory wells, we intend to develop a conceptual model of the hydrogeologic system. Ultimately, this will be used as the building block for the numerical model with which we will attempt to address regional issues raised by the ASR Issue Team and/or CROGEE.

- If the results at sites that have been selected for exploratory wells do not provide the information necessary to constrain a regional hydrogeologic conceptual model, will there be funds available to construct additional exploratory wells in other locations?

To clarify, the data from the 3 monitor wells and the 3 exploratory wells are insufficient in and of themselves to constrain a regional hydrogeologic conceptual model. These new data must be used in conjunction with the regional hydrogeologic information to develop the conceptual model. As stated above, there are insufficient funds in the pilot projects at this time to install additional monitor wells apart from those previously mentioned to fill regional data gaps. Of course, it is difficult to predict the myriad of circumstances that might change this current plan. Note that the CERP process goes to painstaking lengths to allow for mid-course corrections to the plan based on newly acquired information, so we cannot by definition rule this task out. Realistically, project elements other than hydrogeologic investigations (e.g., regulatory, design, construction, testing) currently constrain the pilot project funds available for additional wells above those mentioned herein. The newly proposed Regional Study may allocate funds for additional wells if required.

- Who will undertake the task of developing the regional flow model and running simulations to assess regional effects? Are those who will be involved in model development going to be involved in the planning of data collection activities in order to assure that the necessary data are available to constrain conceptual and numerical models?

At this time, it is envisioned that the USACE will take the lead on implementing the regional groundwater model, with assistance from the SFWMD and the other members of the PDT. USACE options to implement this task include conducting these services in-house, working with the USGS, or out-sourcing the work to architectural-engineering (A/E) firms (as mandated for 40% of their work by Congress). Yes, the model developers will have early input to the data collection activities to ensure that model data needs are met.

Topic II - Water quality issues, regional and local (Note that this topic includes Issue Team items 1, 5 and 6)

Questions that are likely to arise in some form:

- Of the possibilities noted in the various planning documents, which of the proposed sources of water for injection in the Lake Okeechobee and Hillsboro ASR systems are believe to be the most likely?

For the Lake Okeechobee ASR Pilot Project, certainly the Lake itself in conjunction with the major tributaries/estuaries (e.g., Caloosahatchee, Kissimmee, and St. Lucie Rivers) are the most likely sources, given our previous response. For the Western Hillsboro ASR Pilot Project, the Hillsboro Canal and horizontal wells adjacent to the proposed pilot impoundment (to capture seepage) are the most likely sources.

- How will the regulatory concerns associated with injection of these source waters be addressed in the pilot studies?

As the pilot project PMP indicates, two primary tasks are geared towards addressing regulatory issues. First, the source-water quality characterization program attempts to characterize spatial and temporal variability of source water prior to ASR recharge. This is required to obtain an Underground Injection Control (UIC) permit. In addition, it will help determine the parameter list for recovered water. Second, a study to evaluate the fate of microorganisms is included to evaluate the potential for in-situ aquifer conditions to neutralize these microorganisms. It is hoped that results of this study will indicate that microorganisms die-off within the aquifer, precluding the need for expensive pre-treatment (i.e., ultrafiltration) that could result in a project savings of between $250 to $400 million for CERP ASR.

- What is known and what needs to be learned about source water quality, including possible temporal variations?

Basic information about source (surface) water is available, but not to the level required for permits, nor an in-depth understanding of potential geochemical reactions that might occur when it is mixed with aquifer water. As you might expect, much of the data are focused on the phosphorous issue. In general, surface water such as Lake Okeechobee and its tributaries/estuaries has a high organic content. Western portions of the Lake appear to have the best water quality. In addition, there appears to be an inverse relationship between turbidity and

algae content, both potentially problematic issues with respect to potential plugging of ASR zones.

The USGS and SFWMD entered into a cooperative program last year to begin characterizing surface and aquifer water quality, focusing on parameters useful for geochemical modeling purposes.

The proposed source-water quality characterization program attempts to fully characterize water quality and understand temporal variability by conducting sampling seasonally for at least one year, and storm event sampling. Pre-treatment systems must be designed to accommodate temporal changes in water quality, especially during storm events when presumably the greatest volumes of water are available.

- Which of the parameters in the water quality list are important to address regulatory concerns? Which of the parameters in the water quality list are important to address anticipated differences between source waters and ambient groundwater in potential injection zones?

Obviously, the primary and secondary drinking water standards (DWS) are by definition geared towards regulatory concerns, though not exclusively. In addition, some parameters not on the DWS list (e.g., microorganisms) are focused on regulatory issues. Parameters focused on mixing and water-rock interactions include basic cations/anions, chlorides, total dissolved solids, pH, arsenic, radionuclides, and isotopes.

- What is known and what needs to be learned about ambient water quality in potential injection zones, including spatial variations in water quality over the region that will be affected by the ASR systems?

Various hydrogeologic reports attempt to characterize the spatial variability of groundwater within the Floridan aquifer system (FAS). This information will be compiled as part of the hydrogeologic literature search. In addition, limited water quality sampling and analysis of existing FAS wells for basic parameters – principally chlorides, pH, TDS, conductivity, etc. – are conducted on select wells as part of FDEP's ambient water quality monitoring program, partially implemented by SFWMD. Finally, ambient water quality will also be sampled and analyzed from those FAS monitor wells and exploratory ASR wells installed as part of the pilot projects.

In general, ambient water quality appears to freshen as one proceeds northward within a given hydrogeologic unit, ultimately becoming fresh in the Orlando area. Vertically, water quality gradually deteriorates downward to approximately 1,800 feet in South Florida, where a saltwater interface is encountered. There have been documented exceptions to this rule within zones of the upper FAS, but always at depths above 1,800 feet.

- Will the pilot studies provide the necessary data to adequately characterize source water quality and ambient water quality in the injection zones, including temporal and spatial variability?

Yes, we believe that the background- and source-water quality characterization developed will sufficiently characterize these waters to facilitate the pilot projects, as well provide information for regional evaluation of the proposed full-scale ASR implementation.

- If the pilot studies will not provide all the data that may be needed, what additional studies are planned to fill the gaps?

Development of the Regional Study scope of work might identify other studies, but this has not been conducted to date.

- What is known and what needs to be learned about potential water quality changes induced by mixing of source waters and ambient groundwater in the injection zones?

The Florida Geological Survey (FGS) has conducted studies to indicate that leaching of arsenic and radionuclides from the carbonate aquifer matrix has occurred in the Tampa Bay area, relatively close to the Lake Okeechobee ASR Pilot Projects. It is postulated that the influx of highly oxygenated source water within the anaerobic environment of the FAS results in a redox potential conducive to leaching of these bound constituents into solution. Some evidence suggests that this phenomenon decreases over time, but more study on this issue is needed. We intend to rely on the Tampa experiences (and other existing ASR sites, if applicable) to better understand this issue. Given the timeframe to permit/design/construct the ASR Pilot Facilities, we must rely on existing ASR facilities to answer some of these questions near term.

On a positive note, many ASR facilities have been operating in the state of Florida – indeed, globally -- with no observed water quality concerns other than those mentioned above. ASR sites with water quality issues have been observed in granular aquifers, where the potential for plugging due to formation and precipitation of iron oxyhydroxides and manganese exists. Fortunately, the carbonate, solution-riddled aquifers that we propose for ASR storage have been shown to result in little concern for plugging if operated and maintained properly.

- What is known and what needs to be learned about potential water quality changes that may occur due to water-rock interactions?

See previous comment

- Will the geochemical modeling proposed as part of the pilot studies be adequate to predict water quality changes due to mixing and water-rock interactions? Are the necessary thermodynamic and kinetic data for the geochemical modeling readily available, especially for minor, trace, and organic constituents?

Yes, we believe our approach adequately addresses the question about mixing and water-rock interactions. The entity responsible for geochemical modeling will ensure that data needs are identified up front to facilitate geochemical modeling. Note that the USGS and SFWMD previously entered into a cooperative agreement to collect background information that should facilitate geochemical modeling.

Note also that geochemical modeling was conducted as part of SFWMD's ASR demonstration project (CH2M HILL, 1989) near Taylor Creek/Nubbin Slough on the north side of the Lake, which predicted no significant adverse chemical reactions that would preclude ASR storage.

- Will the cycle testing proposed as part of the pilot studies be conducted over appropriate time scales and using appropriate water sources to provide a test of the predictions of the geochemical modeling?

Yes, though cycle testing may not be completed in time for decisions on whether to proceed with the next phase of ASR implementation on a regional scale, as contained in the Project Implementation Report (PIR) phase of the project.

- If observations of water quality changes during cycle testing do not match predictions of the geochemical modeling, will additional studies be undertaken to resolve discrepancies?

Yes, though this will occur during the Project Implementation Report (PIR) phase, following completion of the ASR Pilot Projects.

- What are the direct receptors (e.g. Lake Okeechobee, municipal water supply, canals) and indirect receptors (e.g. water conservation areas and portions of the Everglades that receive flow from the Lake Okeechobee area) for water recovered from the ASR systems at Lake Okeechobee and Hillsboro?

The example receptors embedded in your question are the primary receptors.

- What is known and what needs to be learned about water quality requirements of the receptors for water recovered from the ASR systems?

From a permitting perspective, the environmental requirements include complying the applicable Class I or Class III water quality criteria from NPDES permits. From a municipal perspective, water recovered from ASR systems to Lake Okeechobee must meet Class I (i.e., drinking water) criteria, and also not result in water quality changes such that the water cannot be treated at municipal water treatment plants that serve Glades-area residents.

- Will the pilot studies provide the necessary information to evaluate the suitability of recovered water for its intended receptors?

Yes, a sampling and analysis program for water recovered from ASR systems will be developed and implemented to address these concerns. Note that the details of this plan necessarily require that we await the results of the source-water quality characterization program to ensure that parameters of concern are addressed.

- If the pilot studies will not provide the information needed to assess suitability of recovered water for the intended receptors, what other studies are planned to fill the gaps?

Development of the Regional Study scope of work might identify other studies, but this has not been conducted to date.

> *Topic III – Local performance/feasibility issues (Note that this topic includes Issue Team items 3, 6 and 7)*

Questions that are likely to arise in some form:

- What is known and what needs to be learned about rock properties in potential storage zones and adjacent confining units in order to estimate critical fracture pressures?

Appendix C: Questions to the SFWMD and its Responses 37

Some data exists about rock properties from cores analyzed at other well sites. In addition, preliminary review of rock fracturing literature review indicates that approximately one psi of pressure per foot of overburden has been empirically shown to be the critical pressure for rock fracturing. Assuming a depth of 1,000 feet, it is doubtful that we would generate 1,000 psi of pressure to result in fracturing of the overlying Hawthorn confining unit.

- Will the pilot studies provide the necessary data on rock properties for this analysis? If not, what additional studies are planned to fill the gaps?

Cores will be obtained from the overlying confining units and from the ASR storage zone for laboratory analysis and determination of petrophysical properties.

- Will a regional scale flow model of ASR system operation be adequate to predict local pressure buildups in the vicinity of injection wells for the purpose of evaluating fracture potential? If not, what other types of modeling will be required?

The current version of the pilot project PMP outlines a plan to conduct a desktop analysis using analytical techniques perfected in the oil industry to evaluate induced fracture potential from the proposed regional ASR system. The regional modeling effort should yield pressure information which can be compared with empirical data from the literature (and laboratory analysis of cores) to further evaluate induced fracture potential. If there are other techniques that CROGEE is aware of, we'd be happy to consider their use.

- What is known and what needs to be learned about the potential to inject water in the ASR system at the rates anticipated by the Restudy?

The 5 million gallon per day (mgd) capacity has been demonstrated at two sites (i.e., Miami-Dade Water and Sewer Department West and Southwest ASR Wellfields). Other sites (i.e., West Palm Beach, Western Hillsboro, and Broward County) have demonstrated the hydraulic capacity to recharge water at these rates. We need to learn if the Lake Okeechobee area has an ASR zone with a hydraulic capacity to recharge/recover at this rate. The 3 large-diameter exploratory wells are the first step in addressing this question.

From the broader perspective, admittedly little is known about the ability of the FAS to recharge/recover cumulative flow rates (1,665 mgd) proposed in the CERP. A combination of ground-truthing the 5-mgd per-well capacity in conjunction with the regional groundwater modeling should help answer this important question. More philosophically, 50 years ago it might also have seemed unlikely that the FAS could have supported current withdrawals, although admittedly with environmental and resource implications.

- What is known and what needs to be learned about the potential to recover the required volumes of water with suitable quality for the intended receptors?

Little is known about the ability to of the proposed regional ASR systems to recover the volumes of water needed to meet future environmental, urban, and agricultural demands of South Florida, hence the need to conduct pilot studies. Assuming successful results of the pilot projects, and subsequent step-wise implementation of regional ASR, more will be learned to evaluate the ability of regional ASR systems to meet water quality and quantity needs of intended receptors/users.

- How is "efficiency" of recovery going to be quantified in the pilot studies? Are different definitions of "efficiency" required depending on the intended receptors of the water?

Recovery efficiency is a term open to interpretation, depending on the needs of the end user. Traditionally, it is defined as the amount of water recovered – given a pre-determined upper water quality limit above which water will not be recovered – divided by the volume of water stored. The recovery efficiency term can be skewed because the amount of water left in the zone has an effect of the recovery efficiency of the next cycle. While we will collect data to evaluate recovery efficiency, we prefer to use the term recoverability; that is, is the water available when we need it to meet demands. If so, the recovery efficiency term becomes less important.

- Will the cycle testing conducted in the pilot studies be at appropriate rates and cover appropriate times scales to allow extrapolation of results for the purpose of estimating the recovery efficiency for long term operation under the conditions anticipated in the Restudy?

Cycle testing will be conducted to evaluate appropriate recovery rates. Extrapolation of recovery efficiencies may be more problematic if the traditional approach of building a target storage volume (TSV) over successive cycles is employed. Some evidence suggests that the well-known increase in recovery efficiency over successive cycles can be replicated by storing available water early in the testing program, with little water recovered to establish the subject TSV. It is recommended that when we get to the testing phase, we explore this technique given the experiences at other operational ASR sites.

Topic IV – Other issues, left-over issues, "open mike"

Questions that are likely to arise in some form:

- What additional problems, besides those of the ASR Issue Team, have been identified by the SFWMD to date?

The study to evaluate the fate of microorganisms in aquifers mentioned previously was not specifically raised by the ASR Issue Team, but looms large with respect to the permittability of the proposed ASR systems.

- Do the pilot projects address any of these additional problems?

Yes, a study to evaluate the fate of microorganisms is one of the first tasks to be conducted.

- What are the principles that govern your proposed distribution of funding, and timing, of regional studies vs. site testing?

Good question. The CERP document (April 1999) does not provide many details about the ASR Pilot Projects, an artifact of preparing a typical 5-year feasibility study within the 2-year timeframe mandated by Congress. Having said that, the $19 million figure for Lake Okeechobee $8 million figure for Western Hillsboro, and $6 million figure for Caloosahatchee were crudely based on estimates for a given number of ASR systems, assuming approximately $2 million per system. Unfortunately, there were no specific funding streams to answer some of the regional questions raised by the ASR Issue Team (July 1999, note the date for CERP document). In the

current version of the PMPs, we've attempted to strike the appropriate balance between permitting/designing/ constructing/testing functional ASR systems, and answering regional questions. We think we've achieved that daunting task with our current plan, though the proposed Regional Study will now result in additional funds to conduct regional investigations.

Two underlying themes emerged from our participation in the ASR Issue Team not documented in their report. First, locating all ASR wells in one geographic location (e.g., the west side of Lake Okeechobee) does not answer a fundamental question; that is, does ASR work in geographically dispersed areas around the Lake. The corollary is, it would be difficult to recommend the next phase of ASR implementation (e.g., a 30-well system) in a geographic area that doesn't have a pilot facility in place and successfully operating.

The second theme was, "We know that single-well systems work, we want to know how multiple wells interact with each other." This begins to answer the question, "What is the optimum spacing between ASR wells so that coalesced "bubbles" of freshwater result in an underground reservoir of fresh water."

These three paragraphs are a long-winded justification for six (6) ASR wells, three in geographically dispersed locations and three at an individual ASR "cluster" for the Lake Okeechobee ASR system. Budgetary constraints may force us into a 5-well system for this site, requiring that two wells be located at one of the 3 sites to form the cluster in that fashion. For Western Hillsboro, where the site is defined, only the cluster question need be answered, and it is proposed with a 3-well system, one of which has already been constructed via an existing SFWMD research contract.

Appendix D

Workshop-related materials received by the committee after the workshop and prior to finalization of the report

1. Review of second draft of Project Management Plans for the Lake Okeechobee and Hillsboro Aquifer and Storage Pilot Projects (Robert Renken)
2. USGS response to Subcommittee ASR questions (Robert Renken)
3. Comments of Mark Pearce
4. Comments of Joan Browder
5. Comments of David Pyne
6. Comments of Walt Schmidt
7. Comments of Jim Cowart
8. Additional information about ASR (David Pyne)
9. Evapotranspiration: The Forgotten Key to Everglades Restoration (Juanita Greene)
10. Comments of Michael Waldon

Appendix E

Excerpts from Draft Project Management Plan – Lake Okeechobee

(2nd Draft, September 2000, South Florida Water Management District)

This appendix has been provided for the convenience of the reader, but the CROGEE has made no major editorial changes in the original text as written by the South Florida Water Management District. Minor editorial changes with respect to figures have been made for consistency with the rest of the report, and are noted [*in italics within square brackets*].

SECTION 1
PROJECT INFORMATION

1.1   Description
The *Central and Southern Florida Project Comprehensive Review Study* (USACE, 1999) -- developed jointly by the South Florida Water Management District (SFWMD) and the U.S. Army Corps of Engineers (USACE) – presents a framework for Everglades restoration. Now known as the Comprehensive Everglades Restoration Plan (CERP), this plan contained 68 components, including critical restoration projects, operational changes to the Central and Southern Florida Project (C&SF), creation of water quality treatment facilities and other modifications with the principal goal of the creation of approximately 217,000 acres of new reservoirs and wetlands-based water treatment areas. The CERP achieves the restoration of more natural flows of water, including sheet flow, improved water quality, and more natural hydro-periods in the south Florida ecosystem. Improvements to native flora and fauna, including threatened and endangered species, will occur as a result of the restoration of the hydrologic conditions. The plan was also designed to enlarge the region's supply of fresh water and to improve how water is delivered to the natural system.

A large number of the construction features contained in the CERP were designed at various levels of detail based on information that was available during the plan formulation and evaluation phase. Many of the design assumptions for the components were based solely on output from the South Florida Water Management Model, which averages hydrologic conditions across a model comprised of grid cells with lengths and widths of 2 miles by 2 miles. Consequently, the engineering details of the construction features, including the size and locations are conceptual. More site-specific analyses of the individual components would be needed during the preconstruction engineering and design phase to determine the optimum size, location, and configuration of the facilities. To this end, the CERP contained a number of pilot projects, with the intention of acquiring more information.

Some of the pilot projects described in the CERP include the construction of aquifer storage and recovery (ASR) systems along the Hillsboro Canal, the Caloosahatchee River and adjacent to Lake Okeechobee. This document contains the Project Management Plan to implement the Lake Okeechobee ASR Pilot Project. Figure 1 [*Figure 2 of Chapter 1 of this report*] is a project location map.

The project concept is to store partially-treated surface water or groundwater when it is available in ASR wells – completed within the underlying Floridan Aquifer System (FAS)—for subsequent recovery during dry periods. Among other benefits, implementation of regional ASR technology at the Lake Okeechobee site would help to minimize high-volume water releases to the St. Lucie and Caloosahatchee River estuaries. During dry periods, water recovered from the ASR wells would be used to maintain the surface water level within the lake and associated canals throughout the Everglades, and to augment water supply demands.

The Lake Okeechobee ASR pilot project will consist of six (6) ASR wells, each with an estimated capacity of 5 million gallons per day (mgd). Three (3) of the wells will be located around Lake Okeechobee to demonstrate ASR performance in geographically dispersed areas. A three (3) well cluster will also be installed, to demonstrate how multiple-well ASR systems interact. Monitoring wells and surface facilities will also be constructed at each of these systems. Later phases of the project will include the installation of additional, larger well clusters ultimately reaching the final estimate of 200 ASR wells with the capacity to store and withdraw up to 1 billion gallons of water per day. The purpose of this document is to describe the work tasks that will be necessary to implement the Lake Okeechobee ASR pilot project.

This document provides a comprehensive project management plan for implementation of the Lake Okeechobee Aquifer Storage and Recovery (ASR) Pilot Project through the completion of construction, cycle testing and monitoring. The guidance contained within this document is not intended to be all-inclusive nor to anticipate or include all possible changes to the project during its continuing development. Rather, it is intended to be general in nature as it is expected to be modified, updated and evolve over the life cycle of the project.

1.3 Project Background
Lake Okeechobee lies 30 miles west of the Atlantic Ocean and 60 miles east of the Gulf of Mexico, in the central part of the Florida peninsula. The Lake itself is approximately 730 square miles, and is the principal natural reservoir in south Florida. Portions of Palm Beach, Martin, Okeechobee, Glades and Hendry Counties surround the Lake. Water flows into the Lake primarily from the Kissimmee River, Fisheating Creek and Taylor Creek. Water flows out of the west side of the Lake from the Caloosahatchee River and out of the east side from the St. Lucie and West Palm Beach Canals. The Hillsboro, North New River, and Miami Canals drain the Lake to the south.

Lake Okeechobee is at the center of the south Florida drainage system, receiving flow from the Kissimmee River Basin and to a lesser extent from Everglades Agricultural Area (EAA) backpumping. It discharges east through the St. Lucie (C-44) Canal into the St. Lucie Estuary, west through the Caloosahatchee River (C-43 Canal), and south through four major canals through the EAA into the Water Conservation Areas (WCAs).

In the late 1860s, the Lake was much larger than it is now, with an extensive wetland littoral zone along the shoreline. Water levels fluctuated between 17 feet and 23 feet above National Geodetic Vertical Datum (NGVD), and periodically flooded the exposed areas of the low-gradient marsh. Under both high and low conditions, there was abundant submerged and exposed habitat for fish and other wildlife. Today's Lake is constrained within a dike (i.e., the Herbert Hoover Dike), and the littoral zone is much smaller. As a result, when water levels are above 15 feet NGVD, the entire littoral zone is flooded; leaving no habitat for wildlife that requires exposed ground. When water levels are below 11 feet NGVD, the entire marsh is dry, and not available as habitat for fish and other aquatic life.

Water levels in the Lake are currently regulated by a complex system of pumps, spillways, and locks according to a regulation schedule developed by the USACE. The regulation schedule attempts to achieve multiple-use purposes as well as provide seasonal lake level fluctuations. The schedule is designed to maintain a low lake stage to provide both storage capacity and flood protection for surrounding areas during the wet season. The schedule is also a guide for the management of high lake stages that might threaten the integrity of the Herbert Hoover Dike and thereby risk flooding of downstream lands. During the winter, lake water levels may be increased to store water for the upcoming dry season. This is facilitated by holding water that flows into the Lake from the Kissimmee River Basin and by backpumping from the EAA.

Water quality data indicate that the Lake is currently in a eutrophic condition, primarily due to excessive nutrient loads from the agricultural sources both north and south of the Lake. In the late 1960s and early 1970s, total phosphorus concentrations as low as 50 parts per billion (ppb) were measured. Currently, total phosphorus concentrations in the Lake have been measured in the 100 ppb range. It is likely that historic in-lake turbidity was much lower than current conditions as well.

The CERP presents a new operational plan for the Lake that maximizes water storage opportunities, enhances wildlife populations, restores the ecological health of the Lake, and protects coastal estuaries and public health. ASR technology provides storage - an important component that will contribute to the overall Everglades restoration. The CERP includes the construction of up to 200 ASR wells (with associated pre- and post- treatment facilities) installed adjacent to Lake Okeechobee, with a total combined pumping capacity of 1 billion gallons of water per day. Specifically, the CERP states:

"The purpose of this feature is to: (1) provide additional regional storage while reducing both evaporation losses and the amount of land removed from current land use (e.g. agriculture) that would normally be associated with construction and operation of above-ground storage reservoirs; (2) increase the Lake's water storage capability to better meet regional water supply demands for agriculture, Lower East Coast urban areas, and the Everglades; (3) manage a portion of regulatory releases from the Lake primarily to improve Everglades hydropatterns and to meet regulatory discharges to the St. Lucie and Caloosahatchee Estuaries; and (5) [sic] maintain and enhance the existing level of flood protection."

ASR technology is proposed as a significant storage component in the CERP, with the FAS acting as a large underground reservoir. The advantages of using ASR technology for these objectives include:

- Reduced costs compared with expensive, surface storage facilities
- Eliminates detrimental discharges to the St. Lucie and Caloosahatchee estuaries
- Nearly unlimited underground storage capacity eliminates water losses due to evapotranspiration and seepage
- Wells can be located in areas of greatest need, reducing water distribution costs
- Requires limited land acquisition
- Provides the ability to recover large volumes of water during severe droughts, presumably when reservoir levels would be low

These advantages are particularly important in South Florida where land acquisition costs are high, the availability of water is seasonal, and the underlying FAS are geographically extensive.

The South Florida Ecosystem Restoration Task Force Working Group commissioned the development of an ASR Issue Team in 1998. The Team – consisting of members of the SFWMD, USACE, EPA, FDEP, USGS, other local governmental agencies, and private consultants – met to address the technical and regulatory uncertainties associated with the ASR technology and the scale at which it is proposed in the CERP. The ASR Issue Team identified seven (7) issues that should be addressed prior to full-scale implementation, as presented in their report (ASR Issue Team, July 1999). At least three (3) of these issues are regional in nature. While the ASR pilot projects themselves – including the subject Lake Okeechobee ASR Pilot Project – will not address all seven issues, they will provide valuable site-specific data, which can be used in the regional analyses (including model development) to address all seven issues. A more detailed discussion of the Issue Team items is contained in a subsequent section of this document.

The primary area of investigation for ASR implementation is the perimeter area around the northern rim of Lake Okeechobee, from the City of Moore Haven on the west to the City of Okeechobee to the north, to Port Mayaca on the east, as shown on Figure 2 [*renumbered as Figure E-1 in this report*]. The known occurrence of poorer quality water on the south side of the Lake due to EAA backpumping operations suggests that the remaining area along the Lake perimeter would be a secondary area of investigation.

The area of investigation adjacent to the Lake is characterized as lowland, at an elevation of between 10 to 20 feet above NGVD. Land use is primarily unimproved and improved pasture, wetlands, and occasional areas of planted field crops. State Road 441 runs along the northeast perimeter of the Lake, whereas State Road 78 runs along the northwest perimeter of the Lake.

The final locations and layout of the pilot ASR wells and monitor wells have not been determined at this time. Well siting must incorporate information regarding the proposed footprints of the pilot facilities and feasible conveyance systems to surface water bodies. Well locations and spacing will also be determined based on specific-capacity and aquifer test data obtained from the initial exploratory wells. For more information on the geologic and hydrogeologic setting of the Lake and the aquifers relevant to this project, the reader is referred to Section 4.

Appendix E: Excerpts from Draft Project Management Plan—Lake Okeechobee

Figure 2. Lake Okeechobee Vicinity Map

Figure E-1

WORK BREAKDOWN STRUCTURE

Phase 1 – Initial Project Evaluations

The purpose of Phase 1 is to conduct a preliminary siting evaluation, hydrogeologic analyses, perform a permitting analysis, conduct a study to evaluate the fate of microorganisms in brackish aquifers, and to perform an engineering evaluation of alternatives for constructing the pilot project. The findings of the analyses performed during this phase of work will be included in the Pilot Project Design Report, which will be described in greater detail in a subsequent section of this document. Phase 1 activities are subdivided into the following tasks:

Task 1.1 – Land availability/preliminary siting
        Subtask 1.1.1. GIS Data Compilation and Review.
        Subtask 1.1.2. Detailed On-Site Evaluations.
Task 1.2 – Regional Hydrogeologic Studies
Task 1.3 – Desktop Fracture Analysis
Task 1.4 – Taylor Creek Well Evaluation
Task 1.5 Permitting Evaluation
Task 1.6 Fate of Microorganisms in Aquifers Study
Task 1.7 Engineering Evaluation

Phase 2 – Project Coordination and Public Outreach

Task 2.1 Design Coordination Team Meetings
Task 2.2 Project Delivery Team Meetings
Task 2.3 Independent Technical Review Team and CROGEE Meetings
Task 2.4 Public Outreach Meetings
Task 2.5 PMP Updates and Revisions
Task 2.6 Budget Updates/Revisions

Phase 3 – Surface Water Studies

This project phase focuses on the quality of the surface water in the Lake and its tributaries, and the quality and quantity of water that will be extracted from the ASR wells during recovery. An analysis will be also performed to assess the desired quality of water recovered from the systems, in terms of the requirements of the end users. In addition, pre- and post-treatment alternatives will be evaluated to determine the most cost-effective treatment technology given the source- and receiving water quality, and permitting requirements. The findings of the analyses performed during this phase of work will be included in the Pilot Project Design Report, which will be described in greater detail in a subsequent section.

Task 3.1 – Surface Water Availability Analysis
Task 3.2 – Source Water and Receiving Water Quality Analysis
Task 3.3 - Recovered Water Quality and Quantity Considerations
Task 3.4 – Treatment Alternatives Pilot Testing

Phase 4 - Groundwater Modeling

Groundwater modeling will primarily be conducted to evaluate regional effects of the proposed full-scale CERP ASR implementation on the FAS. Geochemical modeling will be conducted to evaluate the potential for adverse geochemical reactions that could impair ASR performance. Phase 4 can be subdivided into the following tasks:

Task 4.1 – Data Acquisition
Task 4.2 – Conceptual Model Development
Task 4.3 – Finalized Conceptual Model, Computer Code Selection
Task 4.4 – Model Setup
Task 4.5 – Model Calibration
Task 4.6 – Predictive Simulations
Task 4.7 - Model Documentation Report
Task 4.8 – Geochemical Modeling

Phase 5 - Exploratory Wells

To obtain site-specific hydrogeologic information on the nature of the storage zone and FAS water quality, one exploratory well will be permitted, constructed and tested at each of three sites, corresponding to the confluence of major tributaries/estuaries with Lake Okeechobee. At this time, the three proposed locations for the exploratory wells are at the Town of Moore Haven (Caloosahatchee River), Kissimmee River, and Port Mayaca (St. Lucie Canal). The exploratory wells will be constructed with casings large enough to accommodate later installation of permanent pumping facilities and conversion via permitting to operational ASR systems. The activities anticipated during this phase of work are subdivided into the following tasks:

Task 5.1 – Well Design
Task 5.2 – Construction Permitting
Task 5.3 – Contractor Selection and Procurement
Task 5.4 – Well Construction and Testing
Task 5.5 Hydrogeologic/Engineering Report

Phase 6 - Project Development and Documentation

Task 6.1 Prepare Project Management Plan

Task 6.2 Problem Identification

Appendix E: Excerpts from Draft Project Management Plan—Lake Okeechobee 45

Task 6.3 Identification of Objectives and Constraints
Task 6.4 Pilot Project Design Report
Task 6.5 NEPA Compliance
Task 6.6 Pilot Project Design Report
Task 6.7 Project Cooperation Agreement
Task 6.8 Statute 1501 Process Documents
Task 6.9 Pilot Project Technical Data Report

Phase 7 - Well Re-Permitting and Construction

Upon completion of the exploratory well program and development of a strategy to permit the wells for operational use, it is anticipated that applications to convert the well classifications and construct surface facilities to inject into the wells will be filed with FDEP. This phase includes work tasks to perform the permit conversions and to design and construct the required new facilities.

This phase also includes the continued expansion of the pilot project, to include permitting and construction of a 3-well Cluster ASR facility at a new site, FAS and SAS monitor wells and surface facilities at all of the sites. The location of the new cluster facility will have been determined based on the results of the Task 1 Siting phase. Spacing of the wells at the new cluster facility will be based on the results of the exploratory well testing program. The anticipated work tasks included in this phase are:

Task 7.1 – Consultant Selection for Design/Construction Management
Task 7.2 – Final Surface Facility Design
Task 7.3. Water Use Permitting
Task 7.4. NPDES Permitting
Task 7.5. Exploratory Well Re-Permitting
Task 7.6 - Cluster Well Construction Permitting
Task 7.7 – Contractor Selection and Procurement
Task 7.8 – Well Construction
Task 7.9 – Hydrogeologic/Engineering Reports
Task 7.10 – Surface Facility Construction
Task 7.11 – Operation and Maintenance Manuals

Phase 8 - Cycle Testing

Cycle testing will be performed following construction of the exploratory/test wells the multi-well cluster, the FAS and SAS monitor wells, and associated surface facilities, under the UIC Construction Permits for the ASR systems. The purpose is to test the performance of the system, both mechanically and hydrogeologically. Before the system can be tested, however, several permits must be in place.

- Operational Testing Approval from FDEP (tied to the UIC Construction Permits)
- Water Use Permit (for withdrawal from the surface water body)
- WQCE and LAE or Chapter 120 Variance, if applicable (for recharge of partially treated surface water)
- NPDES (for discharge into the Lake, River or Canal)

It will be critical to have these permitting issues resolved in time so that cycle testing can be performed soon after the surface facilities are constructed. After a period of approximately one year of cycle testing, applications for operating the ASR systems on a continuing basis can be filed with the FDEP. Although FDEP has indicated that a minimum of two years of cycle testing data will be required for an operating permit, applications (and supporting documentation) will be prepared after the first year of testing. In this way, other permitting issues can be resolved concurrent with obtaining the second year of data. The tasks included in this phase are:

Task 8.1. Monitoring and Data Collection.
Task 8.2. Reporting During Cycle Testing.

Phase 9 – Operating Permit Applications

Task 9.1 - Operating Permit Applications

Phase 10 – Post Construction Activities

When cycle testing is completed and the operating permits are issued, the pilot project will be complete and final close-out of the project can commence. This section includes the tasks that will be necessary to finish the project. These tasks are:

- Notification of Physical Completion
- Transfer of District Operation and Maintenance Authority
- Final Real Estate Certification and Credits
- USACE Audit
- District Audit
- Prepare Transfer Documents
- Project Fiscal Complete
- Prepare Pilot Project Technical Data Report
- Submit PPTDR To SAD

Appendix F

Excerpts from Draft Project Management Plan – Western Hillsboro

(2nd Draft, September 2000, South Florida Water Management District)

This appendix has been provided for the convenience of the reader, but the CROGEE has made no major editorial changes in the original text as written by the South Florida Water Management District. Minor editorial changes with respect to figures and appendices have been made for consistency with the rest of the report, and are noted [*in italics within square brackets*].

1.0   Project Information

1.1 Introduction

The *Central and Southern Florida Project Comprehensive Review Study (April, 1999)* -- developed jointly by the South Florida Water Management District (SFWMD) and the U.S. Army Corps of Engineers (USACE) – presents a framework for Everglades restoration. Now known as the Comprehensive Everglades Restoration Plan (CERP), this plan contains 68 components, including critical restoration projects, operational changes to the Central and Southern Florida Project (C&SF), creation of water quality treatment facilities and other modifications with the principal goal of the creation of approximately 217,000 acres of new reservoirs and wetlands-based water treatment areas. The CERP achieves the restoration of more natural flows of water, including sheet flow, improved water quality, and more natural hydro-periods in the south Florida ecosystem. Improvements to native flora and fauna, including threatened and endangered species, will occur as a result of the restoration of the hydrologic conditions. The plan was also designed to enlarge the region's supply of fresh water and to improve how water is delivered to the natural system.

The purpose of the Western Hillsboro Aquifer Storage and Recovery (ASR) Pilot Project is to address uncertainties associated with some of the physical facilities that are proposed in the CERP. The pilot project will be designed to determine the feasibility, as well as optimum design, of a facility prior to embarking upon full-scale implementation of the ASR facilities. The formulation of alternative pilot project designs is intended to address cost effective means to address these uncertainties. Formulation of alternative plans for full scale implementation of the Western Hillsboro ASR component will be accomplished during the development of the Western Hillsboro ASR Project Implementation Report, which is scheduled to be initiated upon completion of this pilot project.

A large number of the construction features contained in the CERP were designed at various levels of detail based on information that was available during the plan formulation and evaluation phase. Many of the design assumptions for the components were based solely on output from the South Florida Water Management Model, which averages hydrologic conditions across a model comprised of grid cells with a spacing of 4 square miles. Consequently, the engineering details of the construction features, including the size and locations are conceptual. More site-specific analyses of the individual components are needed during the pre-construction engineering and design phase to determine the optimum size, location, and configuration of the facilities. To this end, the CERP contained a number of pilot projects, with the intention of acquiring more information. One of the pilot projects identified in the CERP is to test ASR technology in the Western Hillsboro Canal Basin (at a location formerly known as Site 1). A map showing the location of the Western Hillsboro site is provided in Figure 1 [*Figure F-1 in this report*] (also Appendix A [*not shown in this report*]).

ASR technology is proposed as a significant storage component at Western Hillsboro and other locations delineated in the CERP. Advantages of ASR include:

- Reduced costs compared with expensive, surface storage facilities
- Underground storage in the upper Floridan Aquifer System (FAS) eliminates water losses due to evapotranspiration and seepage
- Wells can be located in specific areas of greatest need, reducing water distribution costs
- Requires limited land acquisition
- Ability to recover large volumes of water during severe droughts, presumably when reservoir levels would be very low

These advantages are particularly important in south Florida where land acquisition costs are high, the availability of water is seasonal, and the underlying storage zones are expected to be geographically extensive.

The objective of the Western Hillsboro ASR Pilot Project is to test the feasibility of utilizing ASR technology for storage at this site and other sites identified in the CERP. Information gained from the pilot project will be used to develop an operating plan for the system, to refine the long-term operational goals of these and other ASR wells at the site, and to provide insight for future ASR projects, which may be constructed for similar purposes. If it is determined from this pilot project that ASR technology is feasible at the Western Hillsboro site, then the ASR system will be expanded. The expansion will most likely occur in stages, with a total of 30 ASR wells currently proposed at the site. Results from the pilot project will also be useful in determining the feasibility of utilizing ASR technology at other locations, especially within the lower east coast region.

Information to be collected and analyzed during construction and testing of the ASR wells will include:

- Hydrogeologic data from potential storage intervals and confining layers
- Recommendation of the most suitable ASR storage interval(s)
- Quantification of recharge and recovery rates
- Recovery efficiency of the ASR system under differing operating conditions
- Water quality of the source water (surficial aquifer and surface water)
- Water quality of the upper FAS
- Water quality changes in the upper FAS
- Water quality of the recovered water after storage

The full-scale Western Hillsboro Impoundment and ASR system, as outlined in the CERP, will store excess water from the Hillsboro Basin when available (typically in the wet season) and release water into the Hillsboro Canal to maintain canal stages during dry periods.

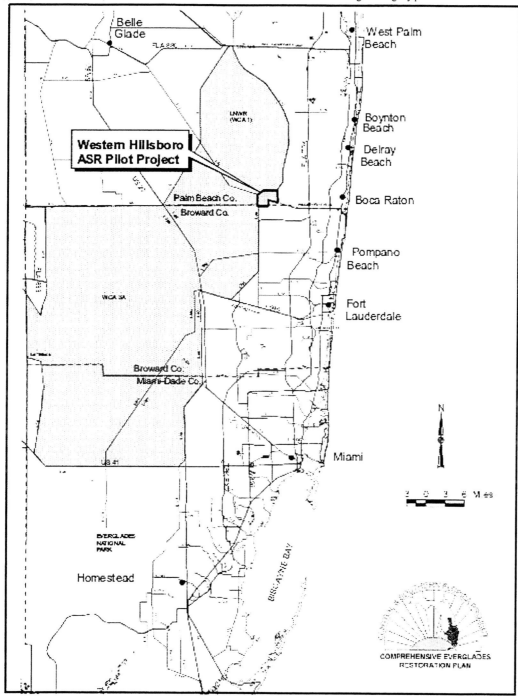

Figure 1.   Project Location Map

1.2 Description

The Western Hillsboro ASR Pilot Project is located along the Hillsboro Canal in southern Palm Beach County. More specifically, it is located at the southern end of the Loxahatchee National Wildlife Refuge (LNWR), also known as Water Conservation Area (WCA) No. 1, where it intersects with the northeastern corner of WCA No. 2 (Figure [F-]1). The 1,660-acre tract north of the Hillsboro Canal is SFWMD-owned, and has been used for improved pasture, nursery stock and aggregate mining. The majority of the site, with the exception of the western-most corner, is currently leased. The leased activities include cattle grazing, and may be expanded to include outdoor entertainment (e.g., skeet shooting, fly-fishing, etc.).

In the CERP, the Western Hillsboro site is ultimately planned to contain a 2,460-acre surface water reservoir (by addition of 800 acres south of the Hillsboro Canal) as shown in Figure [F-]2. The completed ASR pilot project is anticipated to consist of the following components:

- Three ASR wells completed into the upper FAS, each with an anticipated capacity of 5 million gallons per day (mgd).
- A surface water collection system that will supply water to the ASR system. The source of the surface water (e.g. Hillsboro Canal) will depend on water quality testing.
- A groundwater collection system consisting of horizontal wells constructed along the perimeter and underneath the pilot reservoir. The horizontal wells will supply raw groundwater from the Surficial Aquifer System (SAS) to the ASR system.
- A 50-acre pilot impoundment that will be constructed and tested in conjunction with the horizontal wells.
- A pre-injection water treatment facility, if deemed necessary based on water quality testing
- A pre-discharge water treatment facility (for water recovered from the ASR wells, prior to discharge into a canal or reservoir).
- Piping between the source water collection system, the ASR wells, and discharge point(s)
- Surface facilities (e.g., pumps, valves, meters, instrumentation, etc.) to operate and monitor the system.
- Associated monitor wells (FAS and SAS)

A proposed layout of the Western Hillsboro ASR Pilot Project is shown in Figure [F-]3. The layout indicates that shallow horizontal wells, in association with a pilot impoundment, will be used to supply water to the ASR wells for storage.

Surface water may also be used as a source instead of, or in addition to, the horizontal wells. In this case, a surface water collection facility will be located at the most-appropriate location based on water quality. Locations of the ASR wells, monitor wells, and placement of the horizontal wells shown on Figure [F-]3 have been determined based on the objectives of the pilot project. The spacing of the ASR wells will be determined by testing to be completed on the existing exploratory ASR well at the site. Project features shown in Figure [F-]3 also consider the ultimate design of the full-scale impoundment and ASR system. A detailed description of the regional hydrogeology at the Western Hillsboro site is provided in Appendix B [*not shown in this report, but available at http://www.evergladesplan.org/projects/pilot/hillsboro/hillsboro_pp.htm*].

The final design of the Western Hillsboro ASR Pilot Project, to be cost-shared by SFWMD and the USACE, will be based on additional studies and interagency coordination during the preparation of the Pilot Project Design Report (PPDR).

1.4 Background

The Western Hillsboro ASR Pilot Project, as defined by the CERP, consisted of four ASR wells supplied with groundwater from a series of vertical, shallow wells. The framework of the pilot project, however, has changed since the description in the CERP. First, both ground-water and surface-water sources will be evaluated during the pilot project. The second change is that, if groundwater is used, shallow horizontal wells will be used instead of deeper vertical wells, as originally proposed in the CERP. This change was made because of water quality and quantity considerations at the site. In other words, horizontal wells are expected to yield more water and higher quality groundwater (i.e., lower chloride concentration) than the original vertical well concept. Lastly, and as a result of the change to horizontal wells, the planned pilot impoundment becomes integral to the pilot ASR project. This is because the horizontal wells will be installed underneath and along the perimeter of the pilot impoundment. This will allow more of the water in the impoundment to be captured by the horizontal wells for recharge into the ASR wells.

As stated previously, a pilot impoundment will be constructed, tested, and operated in association with the ASR pilot project. The SFWMD recently completed, as part of the Lower East Coast Water Supply Plan, a feasibility study for an impoundment at the site (Montgomery-Watson, 1999) [*no full reference given in original document*]. The results of the study indicated that an impoundment could be constructed at the site, although there would be some expected water losses associated with evaporation and seepage into underlying strata. The SFWMD has made progress on the design of a 50-acre pilot impoundment at the site, which will include a groundwater collection system (horizontal wells) for the pilot ASR system. Test results from the operation of the pilot impoundment and collection system will help determine feasibility and design considerations for a full-scale impoundment and ASR system at the site.

To date, the SFWMD has installed two wells at the western corner of the site. The first is a FAS monitor well – designed to evaluate water level and hydraulic information within zones of the FAS. The second is a 24-inch-diameter exploratory ASR well. The exploratory well was permitted by Florida Department of Environmental Protection's (FDEP's) Underground Injection Control (UIC) program in December 1999. Once the necessary water quality data have been collected, an application may be submitted to FDEP for an ASR construction permit. Figures [F-]4 and [F-]5 present construction completion diagrams for the FAS monitor well and exploratory ASR well, respectively.

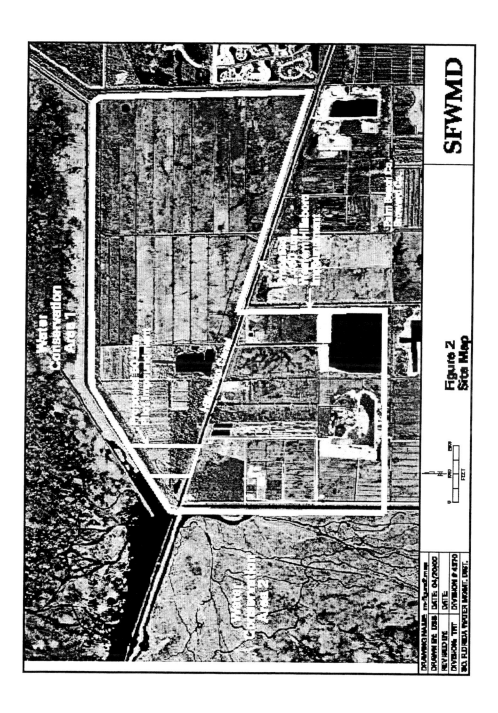

Figure F-2

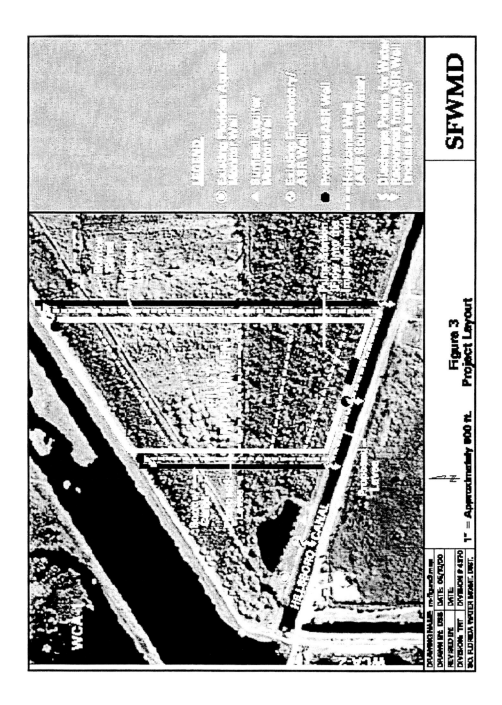

Figure F-3

Appendix F: Excerpts from Draft Project Management Plan—Western Hillsboro

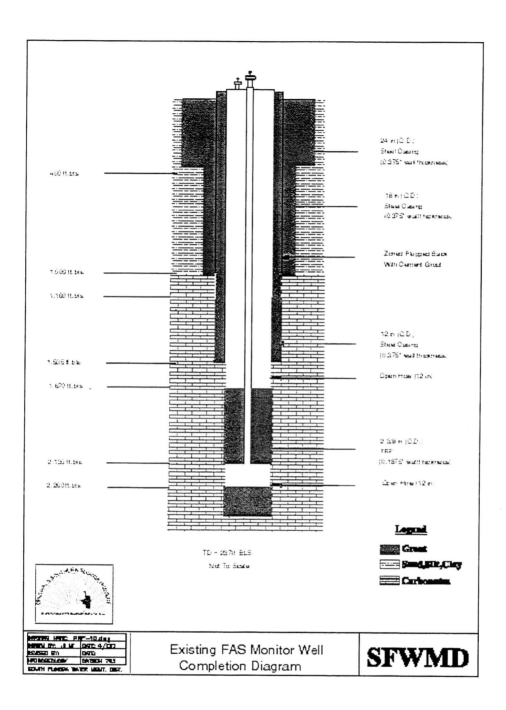

Figure F-4

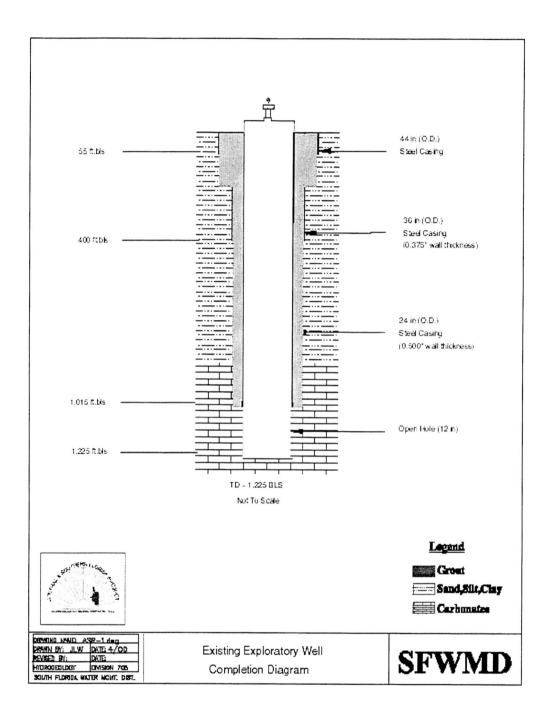

Figure F-5

Appendix F: Excerpts from Draft Project Management Plan—Western Hillsboro 53

1.5 Related Projects

The CERP includes one component that is directly related to this ASR pilot project. This is the full-scale impoundment with ASR at Western Hillsboro. The component, as described in the CERP, includes a 2,460-acre, above-ground impoundment with fluctuating water levels up to 6 feet above grade. The final design of the impoundment will be determined through the Water Preserve Area Feasibility Study. The full-scale component at Western Hillsboro also includes ASR technology with a capacity of injecting and recovering 150 mgd of surplus water from the basin. Groundwater from the surficial aquifer was assumed to be the source of water for storage in the ASR wells.

The Western Hillsboro ASR Pilot Project is directly related to the full-scale component because it will be used to assess the feasibility of using ASR technology at this site, and other sites in the lower east coast region.

Although not a CERP project, the SFWMD is currently co-funding the construction of an ASR well approximately five miles east of the pilot project location along the Hillsboro Canal. Palm Beach County is constructing one ASR well, along with a series of shallow, vertical water supply wells. Once completed, the Eastern Hillsboro ASR well will be used by the County to help meet its customers water demands, in addition to supplying regional benefits to the basin.

2.0   Project Scope

The project scope provides a summary and general description of the tasks to be performed and services to be provided as part of this project. This scope is the basis for the more detailed Work Breakdown Structure, the project schedule, and cost estimate....

Task 1 – Initial Exploratory ASR Well and Monitor Well

This task has been completed with the alteration of an existing SFWMD contract. Two wells were installed under the contract. One well is a FAS monitor well – designed to evaluate water level and hydraulic information within zones of the FAS. The second is a 24-inch-diameter exploratory ASR well. The exploratory well was permitted by FDEP's UIC program in December 1999. The exploratory well construction was completed in June 2000. Testing of the exploratory well (withdrawal only) will be completed in the next several months. The exploratory ASR well will be repermitted as an ASR well and is planned to be one of the three pilot ASR wells at the site. The FAS monitor well will also serve the pilot project in a monitoring capacity.

Task 2 – Project Coordination and Public Outreach

This task involves project coordination between the SFWMD, USACE, and other entities such as the Project Delivery Team that will be involved with the development and execution of this pilot project. This coordination is further explained in Section 5.0 Organization Breakdown Structure. Public involvement will be critical to the success of the pilot project. This will include soliciting the involvement of those parties interested in the development and progress of the project.

Task 3 – Hydrogeologic and Hydrologic Investigations

The primary purpose of this task is to collect existing hydrogeologic and hydrologic information that is relevant to the pilot project. This preliminary investigation will focus on literature review of existing data specific to the study area and the aquifer systems that may be encountered during the project. Information collected during this preliminary literature review will be used as a foundation to build a conceptual hydrogeologic model. This conceptual model will be used to in the formation of the regional groundwater model (Task 4). Information from the literature review will also be used to perform a desktop analysis of potential fracturing of the confining beds that overly the intended ASR storage zones. This fracture analysis will be updated as more information is collected during the construction of the ASR wells. The investigation also includes a water availability and demand analysis for the area.

Task 4 – Modeling

This task includes two types of modeling. First, a numerical, groundwater model will be used to simulate regional effects from ASR activities related to CERP. The regional groundwater model will be built upon the information gathered during the literature review and the conceptual hydrogeologic model development in Task 3. The regional model will be used to simulate such regional effects as changes in potentiometric heads within the FAS, and pressure buildup (and decrease) during ASR activities. Both of these effects were brought up by the ASR Issue Team. The second type of model is a geochemical model. Water samples from the existing FAS wells at the site will be used for the model. The model can be updated once cores from the intended storage zone are collected during the construction of the remaining ASR wells.

Task 5 – Water Quality Characterization and Treatment

This task includes three primary activities related to water quality and treatment. The first is the sampling and analysis of surface water in support of the permitting and design of the ASR pilot system. These surface water bodies include the Loxahatchee National Wildlife Refuge, Water Conservation Area No. 2, and the Hillsboro Canal. The second activity is a microbiological fate analysis. The purpose of the analysis is to evaluate the fate of coliform bacteria and other microorganisms in the subsurface. This is will be an important issue related to permitting a surface water ASR facility at the site. The final activity will be a water treatment analysis. Depending on the quality of the source water, some measure of pre- and/or post-treatment will be necessary for the ASR pilot system

Task 6 – Design, Permit, and Construct Pilot Impoundment and Horizontal Wells

This task includes the design, permitting, and construction of the pilot impoundment and groundwater collection system (horizontal wells). The pilot project currently assumes that either or both surface water and groundwater will be recharged and stored in the FAS. In the CERP, it was assumed that the source water for the ASR pilot project would groundwater from traditional vertical wells. However, because of limitations in the groundwater quality and quantity at the site, horizontal wells in association with an above-ground impoundment were determined to be the optimal source of groundwater.

Task 7 – Test and Operate Pilot Impoundment and Horizontal Wells

This task includes the operation and testing of the pilot impoundment in conjunction with the periodic testing of the horizontal wells. The purpose of this task is to test the effectiveness of the groundwater collection system with regards to water quantity and water quality. Based on the testing results, a determination will be made on the feasibility of using groundwater as source water for the ASR system. The pilot impoundment will be operated continually during this period in support of data collection for the full-scale impoundment proposed at the site.

Task 8 – Project Development Process

This task involves the processes necessary to complete the pilot project. The process starts with the preparation and approval of this Project Management Plan (PMP). The next step is the preparation of the Pilot Project Design Report (PPDR), which includes design and permitting phases of the project. In parallel to the PPDR, (National Environmental Policy Act (NEPA) requirements will be addressed. Before significant construction can begin, the Project Cooperation Agreement (PCA) must be prepared and approved. Also included in the process is the State of Florida's Chapter 373.1501 requirements for state funding. Finally, a Technical Data Report (TDR) will be prepared at the end of the pilot project. The TDR will summarize the findings of the pilot project and make recommendations for further study. Information from the TDR will be used during preparation of the Project Implementation Report (PIR) of the full-scale ASR component of CERP.

Task 9 – Permitting

This task involves the permits necessary to construct and operate the pilot ASR system. Permitting requirements may include Class V Construction Permits, Operating Permits, Water Quality Criteria Exemptions (WQCEs), Limited Aquifer Exemptions (LAEs), an National Pollutant Discharge Elimination System (NPDES) permit, and a Water Use Permit. A FDEP UIC ASR Well Construction Permit will be needed before construction can begin. The UIC construction permit will also allow limited cycle testing (Task 11) of the pilot ASR system. The existing ASR exploratory well will be re-permitted in addition to permitting the remaining two ASR wells. If any secondary drinking water standards are exceeded, a WQCE from FDEP may also be required. An exemption may also be required for the exceedance of primary drinking water standards such as coliform (which is common in surface water).). A LAE is the current regulatory relief mechanism allowed by FDEP if coliform is exceeded in the source water of an ASR system. EPA has also indicated that it may allow a risk-based approach to the exceedance of coliform prior to recharge into an ASR well. These regulatory issues should become clearer as the pilot project progresses. A NPDES permit will be required before cycle testing can begin. The NPDES permit covers the discharge of water recovered from the ASR wells back into a surface water body. The Water Use Permit would be required for the "use" of groundwater or surface water for storage in the ASR wells.

Task 10 – Design and Construct Remaining ASR Wells, Monitor Wells, and Surface Facilities

This task involves the design and construction of the two remaining ASR wells for the pilot project. The first exploratory ASR well was constructed under Task 1. Following the issuance of an ASR construction permit (Task 9), the remaining two ASR wells can be constructed. The construction of the remaining ASR wells will be similar to the existing well. The task also includes the construction of an additional FAS monitor well, and a SAS monitor well. The construction will conclude with the installation of the surface facilities. The surface facilities include the piping, pumps, instrumentation, etc. necessary to operate the ASR pilot system.

Task 11 – Cycle Testing

This task involves the cycle testing of the pilot ASR system. Once all the components of the ASR pilot project are constructed, permitted, and operational (ASR wells, monitor wells, treatment facility, etc.), then cycle testing may begin. This task involves operation, data collection, and reporting necessary to support the evaluation of the pilot project.

Task 12 – Post-Construction Activities

This task involves all those activities that will be necessary to complete before construction is deemed complete. These activities include certifications, credits, audits, and reports necessary to close out the project.

Appendix G

Biographical Sketches of Committee Members

James M. Davidson, Chair, recently retired as vice president for agriculture and natural resources at the University of Florida, a position he had held since 1992. From 1979 to 1992 Davidson served as assistant dean and dean for research for the University of Florida's Institute of Food and Agricultural Sciences (UF/IFAS). He arrived at the UF/IFAS in 1972, as a visiting associate professor and joined the faculty as a soil science professor in 1974. Davidson previously taught at Oklahoma State University and held laboratory research positions at Oregon State University and at the University of California, Davis. A widely recognized expert in hydrology and agronomy, he has served on numerous committees investigating groundwater quality, including the Water Science and Technology Board (1986-1990). He earned a bachelors degree in soil science and a masters degree in soil physics from Oregon State University and a doctorate in soil physics from the University of California, Davis.

Scott W. Nixon, Vice Chair, is professor of oceanography and director of the Rhode Island Sea Grant College Program at the University of Rhode Island. He currently teaches both graduate and undergraduate classes in oceanography and ecology. His current research interests include coastal ecology, with emphasis on estuaries, lagoons, and wetlands. He has served on three National Research Council committees including, most recently, the Committee on Coastal Oceans. Dr. Nixon received a B.A. in biology from the University of Delaware and a Ph.D. in botany/ecology from the University of North Carolina-Chapel Hill.

John S. Adams is professor and chair of the Department of Geography at the University of Minnesota. He researches issues relating to North American cities, urban housing markets and housing policy, and regional economic development in the United States and the former Soviet Union. He has been a National Science Foundation Research Fellow at the Institute of Urban and Regional Development, University of California at Berkeley, and economic geographer in residence at the Bank of America World Headquarters in San Francisco. He was senior Fulbright Lecturer at the Institute for Raumordnung at the Economic University in Vienna and was on the geography faculty of Moscow State University. He has taught at Pennsylvania State University, the University of Washington, and the U.S. Military Academy at West Point. His most recent book, Minneapolis-St. Paul: People, Place, and Public Life, looks at the region's growth and at what factors may affect the metropolitan area's future. Adams holds two degree in economics and a doctorate in urban geography from the University of Minnesota.

Jean M. Bahr is professor the Department of Geology and Geophysics at the University of Wisconsin-Madison where she has been a faculty member since 1987. She served as chair of the Water Resources Management Program, UW Institute for Environmental Studies, from 1995-99 and she is also a member of the Geological Engineering Program faculty. Her current research focuses on the interactions between physical and chemical processes that control mass transport in ground water. She earned a B.A in geology from Yale University and M.S. and Ph.D. degrees in applied earth sciences (hydrogeology) from Stanford University. She has served as a member of the National Research Council's Board on Radioactive Waste Management and several of its committees.

Linda K. Blum is research associate professor in the Department of Environmental Sciences at the University of Virginia. Her current research projects include study of mechanisms controlling bacterial community abundance, productivity, and structure in tidal marsh creeks; impacts of microbial processes on water quality; organic matter accretion in salt marsh sediments; and rhizosphere effects on organic matter decay in anaerobic sediments. Dr. Blum earned a B.S. and M.S. in forestry from Michigan Technological University and a Ph.D. in soil science from Cornell University.

Patrick L. Brezonik is professor of environmental engineering and director of the Water Resources Research Center at the University of Minnesota. Prior to his appointment at the University of Minnesota in the mid-1980's, Dr. Brezonik was professor of water chemistry and environmental science at the University of Florida. His research interests focus on biogeochemical processes in aquatic systems, with special emphasis on the impacts of human activity on water quality and element cycles in lakes. He has served as a member of the National Research Council's Water Science and Technology Board and as a member of several of its committees. He earned a B.S. in chemistry from Marquette University and a M.S. and Ph.D. in water chemistry from the University of Wisconsin-Madison.

Frank W. Davis is a Professor at the University of California Santa Barbara (USCB) with appointments in the Donald Bren School of Environmental Science and Management and the Department of Geography. He received his B.A. in biology from Williams College and a Ph.D. from the Department of Geography and Environmental Engineering at The Johns Hopkins University. He joined the Department of Geography at UCSB in 1983, and established the UCSB Biogeography Laboratory in 1991. His research focuses on the ecology and management of California chaparral and oak woodlands, regional conservation planning, satellite remote sensing of regional land cover, and GIS modeling of species distributions. He was Deputy Director of the National Center for Ecological Analysis and Synthesis between 1995 and 1998, and currently directs the Sierra Nevada Network for Education and Research Page. Dr. Davis has been a member of two prior NRC committees and is currently serving on the NRC Committee on the Second Forum on Biodiversity.

Wayne C. Huber is professor and head of the Department of Civil, Construction, and Environmental Engineering at Oregon State University. Prior to moving to Oregon State in 1991, he served 23 years on the faculty of the Department of Environmental Engineering Sciences at the University of Florida where he engaged in several studies involving the hydrology and water quality of South Florida regions. His technical interests are principally in the areas of surface hydrology, stormwater management, nonpoint source pollution, and transport processes related to water quality. He is one of the original authors of the Environmental Protection Agency's Storm Water Management Model (SWMM) and continues to maintain the model for the EPA. Dr. Huber holds a B.S. in engineering from the California Institute of Technology and an M.S. and Ph.D. in civil engineering from the Massachusetts Institute of Technology. He is currently a member of the NRC's Committee on Causes and Management of Coastal Eutrophication.

Stephen R. Humphrey is dean of the College of Natural Resources and Environment at the University of Florida where he also serves as affiliate professor of Latin American studies, wildlife ecology, and zoology. He also has been the curator in ecology for the Florida Museum of Natural History since 1980. Dr. Humphrey has authored and co-authored numerous articles and books on the effects of urbanization on wildlife. He holds B.A. in biology from Earlham College in Richmond, Indiana and a Ph.D. in zoology from Oklahoma State University. He is chair of the Environmental

Regulatory Commission of the Florida Department of Environmental Regulation and a member of the Florida Panther Technical Advisory Council of the Florida Game Commission.

Daniel P. Loucks is professor of civil and environmental engineering at Cornell University. His research, teaching, and consulting interests are in the application of economics, engineering, and systems theory to problems involving environmental and water resources development and management. Dr. Loucks has held visiting appointments at a number of universities in the US and abroad, worked for the World Bank, the International Institute for Hydraulic and Environmental Engineering, the International Institute for Applied Systems Analysis. He has been a consultant to a variety of government and international organizations on projects involving water resources development and management in Africa, Asia, Eastern and Western Europe and South America. He is a member of the National Academy of Engineering and most recently served on the National Research Council's Committee on Risk-Based Analyses for Flood Damage Reduction Studies.

Gordon H. Orians (NAS) is professor emeritus of zoology at the University of Washington. Dr. Orians began teaching at the University of Washington in 1960 as an assistant professor of zoology and was director of the Institute for Environmental Studies 1976-1986. Dr. Orians has done pioneering research on the evolution of vertebrate social systems, both developing theory and testing a rich assortment of ideas. His research embodies studies on interspecific territoriality and optimal and central place foraging and integrates the concepts of environmental quality and habitat selection. Dr. Orians is a member of many professional societies and academies including the National Academy of Sciences. He has served on numerous National Research Council committees, including his current service as chair of the Board on Environmental Studies and Toxicology. He holds a B.S. from the University of Wisconsin and a Ph.D. from the University of California, Berkeley in zoology. (Resigned from Committee as of Dec. 18, 2000)

Kenneth W. Potter is professor of civil and environmental engineering at the University of Wisconsin-Madison. His expertise is in hydrology and water resources, including hydrologic modeling, estimation of hydrologic risk, estimation of hydrologic budgets, watershed monitoring and assessment, and aquatic ecosystem restoration. He received his B.S. in geology from Louisiana State University and his Ph.D. in geography and environmental engineering from The Johns Hopkins University. He has served as a member of the NRC's Water Science and Technology Board and several of its committees.

Larry Robinson is director of the Environmental Sciences Institute at Florida A&M University where he is also a professor. At Florida A&M University he has led efforts to establish B.S. and Ph.D. programs in environmental science in 1998 and 1999, respectively. His research interests include environmental chemistry and the application of nuclear methods to detect trace elements in environmental matrices and environmental policy and management. Previously he was group leader of a neutron activation analysis laboratory at Oak Ridge National Laboratory (ORNL). At ORNL he served on the National Laboratory Diversity Council and was President of the Oak Ridge Branch of the NAACP. Dr. Robinson earned a B.S. in chemistry, summa cum laude, from Memphis State University and a Ph.D. in nuclear chemistry from Washington University in St. Louis, Missouri.

Steven E. Sanderson is Vice President for Arts and Sciences and Dean of Emory College at Emory University in Atlanta, Georgia. In the mid-1980s, Dr. Sanderson served as Ford Foundation Program Officer for Rural Poverty and Resources in Brazil, where he designed and implemented the foundation's Amazon Program. He served on the faculty of the University of Florida from 1979 to 1997, chairing the Department of Political Science and directing the Tropical Conservation and Development Program. From 1994-97 he chaired the Social Science Research Council Committee for Research on Global Environmental Change. He served on the National Research Council's Committee on Human Dimensions of Global Change from 1993-1996. Dr. Sanderson earned a B.A. in history from the University of Central Arkansas, and an M.A. in political science from the University of Arkansas. He earned a second M.A. and Ph.D. in political science from Stanford University.

Rebecca R. Sharitz is professor of botany at the University of Georgia and senior scientist at the Savannah River Ecology Laboratory in Aiken, South Carolina. Currently, she researches ecological processes in wetlands, including factors affecting the structures and function of bottomlands hardwood and swamp forest ecosystems, responses of wetland communities to environmental disturbances, and effects of land management practices on nearby wetland systems. Dr. Sharitz has served on several NRC committees including, most recently, The Committee on Noneconomic and Economic Value of Biodiversity: Application for Ecosystem Management. She received a B.S. in biology from Roanoke College and a Ph.D. in botany and plant ecology from the University of North Carolina.

John Vecchioli recently retired as a hydrologist with the U.S. Geological Survey's Water Resources Division in Tallahassee, Florida and as chief of the Florida District Program. Previously, he was responsible for quality assurance of all technical aspects of ground water programs in Florida. His research interests have included study of hydraulic and geochemical aspects of waste injection in Florida and of artificial recharge in Long Island, N.Y. He has also done research on ground water-surface water interactions in New Jersey and Florida. Mr. Vecchioli received his B.S. and M.S. in geology from Rutgers University. Mr. Vecchioli previously served on the NRC's Committee on Ground Water Recharge.